T0075162

CURIOUS SPECIES

WHITNEY BARLOW ROBLES

Curious Species

HOW ANIMALS MADE NATURAL HISTORY

Yale UNIVERSITY PRESS NEW HAVEN AND LONDON

Published with assistance from the foundation established in memory of Calvin Chapin of the Class of 1788, Yale College.

Yale University Press books may be purchased in quantity for educational, business, or promotional use. For information, please e-mail sales.press@yale.edu (U.S. office) or sales@yaleup.co.uk (U.K. office).

Set in Scala Pro and Scala Sans Pro type by Integrated Publishing Solutions.
Printed in the United States of America.

ISBN 978-0-300-26618-4 (hardcover : alk. paper)
Library of Congress Control Number: 2023933781
A catalogue record for this book is available from the British Library.

This paper meets the requirements of ANSI/NISO Z39.48-1992 (Permanence of Paper).

10 9 8 7 6 5 4 3 2 1

To all the animals I have loved. Including my husband.

CONTENTS

PREFACE

I have always been curious about other animals, and for as long as I can remember, my interest in them has been regarded as somewhat curious too. As a child, while most little girls played with dolls, I was out in the woods kissing frogs with no expectation of a prince, dangling daddy longlegs in failed hypnosis experiments with my peers, and trying to make a buck selling live insects in lieu of lemonade from a stand at the end of our driveway. The neighbor across the street, in response, crowned me the Queen of Disgust. Fast forward to a time when I saw my career as a professional science writer and editor suddenly morph into the pursuit of a doctorate in early American studies. That first semester, I found myself facing off with a museum specimen of a chain pickerel, an elongated freshwater fish with a distinct underbite and sides mottled like chicken wire. Its skin had been dried, flattened, and slapped onto an annotated piece of paper outside Boston in the late 1700s, near the end of the Age of Enlightenment—a flurry of scientific inquiry, philosophical upheaval, and political revolution stretching across the eighteenth century. Even in the presence of a fish so still, my curiosity about the animals of the past, and about the curious people who studied them, began to quicken.

That shadow of a fish inspired this book. It tells a story about the people and, most of all, the animals who together animated the capacious discipline of natural history in the eighteenth century, the quintessential era of curiosity. Although most books on this early form of natural science stay rooted in history (and, largely, the human drivers of events), *Curious Species* straddles the eighteenth century and the present day. Its chapters weave back and forth between histories of specific creatures—chief among them corals, rattlesnakes, fish, and raccoons—and narratives of my own recent escapades with the descendants of these animals in the twenty-first century. As we will see, modern human-animal relations owe many a debt to the science of three centuries ago. This book reorients our picture of how science gets done and how knowledge gets made, or lost, around who or what gets studied. It shows how our world relies on an ecology of other beings beyond human control or design. It makes animals its unruly protagonists.

At a historical moment when natural history coalesced into a discipline and likewise sought to discipline nature through classification and experiment, animals fundamentally shaped and thwarted the human quest to know the natural world. Corals sank ships while defying categorization; cryptic rattlesnakes so terrified naturalists that some declined to study them; and raccoons, as curious curiosities themselves, muddied what it meant to be human and what it meant to know. These animals reveal a forgotten side of early science that embraced nature's resistance to human probing—an indulgence in ignorance that haunts our inability to understand such beings today. Accessing the elusive lives of historical beasts, their modern relevance, and their pressing ecological stakes beckons a rhetorical intervention into how the history of eighteenth-century science has been told, blending past and present, historian and naturalist. Through its empathic writing style, the story that follows wades into the subjective science of Enlightenment-era naturalists and the unfathomable nature of the creatures they studied.

Given the book's interdisciplinary nature and hybrid form, I have written it with various readers, both scholarly and nonspecialist, in mind. Historians—of science, the early Americas, environmental history, intellectual history, and race and colonialism—will find arguments about the animal foundations of human knowledge, ignorance as a virtue, creaturely effects on colonialism, and yet more evidence for the contributions of enslaved and Indigenous expertise to the early sciences. Scholars of animal studies and material culture can expect to encounter historically rooted examinations of animal agency and power and accounts of the vitality of museum objects. Practicing scientists can learn something of the history behind their own subfields, materials, and methods. And general readers interested in animals, natural history, environmental issues, the history of science, or simply a good adventure story will see how fish paved the way for batteries, why coral made Captain Cook tremble, and whether raccoons might yet become our furry overlords. For all readers, I hope the book can suggest why our global environmental crisis and its many problems of knowledge require learning something about the animals of the past, and why we need to look back as we look ahead.

CURIOUS SPECIES

INTRODUCTION

Animal Matters

TO GATHER CREATURES FROM THE DEPTHS of the sea, a naturalist in the 1700s would cast a wide net—and perhaps enlist the help of a shark. Such assistance was, to be sure, coerced. People sliced open shark bodies to sort through their innards for "divers strange Animals not easily to be met with elsewhere," according to the collecting instructions of the London apothecary James Petiver. He called on his far-flung suppliers to "pray look for, and preserve" hidden jewels in these bellies buzzing with possibility. Centuries before the development of ecology as a field, naturalists uncovered otherwise inaccessible facts by capitalizing on relationships in the natural world, including the knowledge of the deep possessed by a wide-ranging apex predator.

Petiver was one of many naturalists working across the British Empire to amass an archive of animal, plant, and mineral specimens. These would be acquired not just through nonhuman collectors but also, in Petiver's case, connections in the slave trade, a system that brutally reduced people to the status of tradeable goods. Rummaging through the belly of the beast, however, forced naturalists to face a haunting truth: in addition to ocean oddities, they occasionally pulled semi-digested humans out of the stomachs of sharks. Colonial naturalists may have fancied themselves masters of men and the sea in this so-called Age of Sail. Animals showed they could face a fate worse than Jonah's.[1]

In a word, those sharks that furnished specimens and swallowed humans mattered to the making of scientific knowledge. Their appetites mattered, their habits and habitats mattered, their repurposing as a fishing net mattered, their knowledge mattered, and their ability to make people question their control of nature and their place within it—to reconsider what it meant to be human by discovering *themselves* in animals—mattered. Their very matter mattered.

It matters still. Today's naturalists, now highly specialized and bearing titles like ichthyologist and malacologist, don't only use new-fangled technologies and genetic analysis to locate and describe new species. They still get bloody prying open the guts of predators in pursuit of new creaturely finds like their forebears did three centuries ago. Even the specimens used by scientists to define and describe a species writ large have been found using this method of built-in bycatch, despite the depredations of gastric juices. "If your source material is limited—you use what you have," Mark Sabaj told me; he's a scientist who manages the fish collection of Drexel University's Academy of Natural Sciences. Jonah's icefish, formally *Neopagetopsis ionah,* got its title from being found in the belly of a whale. The Latin species name of *Ethadophis merenda,* a fish that has only been documented once, means afternoon snack, since it became just that for a white sea bass. It is a snake eel, but commonly called the snack eel. And in 2001 researchers discovered an Australian snailfish that has never been seen again, at least not by mainstream scientists. The ichthyologists who named the snailfish *Paraliparis infeliciter* found it in the belly of a deep-sea orange roughy. Its species name amounts to bad luck.[2]

This book uses manifold moments of animal influence such as this, moments of striking parallelism between the eighteenth century and now, to show how animals have long shaped the quest for scientific knowledge. It uncovers how humans come to know what they know by investigating how the moving targets of inquiry guide curiosity's path in fundamental ways. The very thing being probed affects what we can see—or what we can't.

Living animals and long-dead specimens were some of the star subjects of study in the intellectual project known as natural history in the eighteenth century. The midpoint of this century saw the birth both of modern museums and modern taxonomy, or the naming and classification of animals and other natural entities. Early scientists encountered and studied a swirling storm of beasts, plants, minerals, and environments through the currents of global exploration and violent colonial spread. In this pulsing world of nature study, animals in particular put humans in a double bind: naturalists were damned if they did and damned if they didn't work with them.

Animals aided the work of natural history in surprising ways, not just as passive bellies but also as moving and feeling participants. They touched every step in the process of making and disseminating knowledge about the natural world, including the collection and preservation of specimens, the journey of specimens to repositories, the writing of an animal's history, the alchemy of turning animals into symbols, and even the choice of visual technique used to represent a particular species on paper. Their influence would reach across what historians call the "long" eighteenth century, blurry on the edges and bleeding into the century before and after. Animals could be cunning collaborators, tools and raw materials for the practice of natural science, and unwitting co-creators of an enterprise that fueled empires.[3]

But animals also thoroughly destabilized that enterprise by overturning the best-laid plans of naturalists and remaining impenetrable to scientific probing. If animals quite literally made natural history, they made its obverse as well. Especially in the case of creatures that humans found unpleasant or terrifying or otherwise hard to read, animals might foreclose the work of understanding altogether. When the English naturalist Daines Barrington contemplated the bat in 1781, for instance, he found it "so disagreeable an animal, that we are generally desirous of avoiding it rather than examining into its habits; the consequence of which aversion is, that we are more ignorant with regard to its natural history." Early science was about gaining knowledge, but also, at times, letting go of it.[4]

In many ways, the world of eighteenth-century natural history seems a distant leap from the one we inhabit today. Look around your living space. Then look at the painting of the Dutch naturalist Laurens Theodoor Gronovius sitting in his study with his two sons, the very picture of an enlightened Enlightenment naturalist (figure I.1 and plate 1). Are there jarred specimens at your disposal? A microscope wrapped in stingray skin on the mantel? What secrets do your cabinets hold? Is there a bezoar, a stone formed in the recesses of an animal's gut, in the drawer? Could you demonstrate for guests the wonders of electricity by cranking a tabletop device? Have you dressed your son in a fantastic gown? (And, if he's still in diapers, ask yourself: why not?) You might explain away those trappings as a matter of privilege and power. Surely, Gronovius was well-to-do, inheriting wealth and expanding upon a collection passed down from his father, the botanist Jan Frederik Gronovius. Many a modern tycoon accumulates; consider, for example, the oligarch Viktor Vekselberg and his collection of Fabergé eggs. But not even Elon Musk, the oddest of the odd billionaires of our own times, has a pickled siren.

Figure I.1. Gronovius and other curious species in his cabinet. Isaac Lodewijk la Fargue van Nieuwland, *Portret van Laurens Theodorus Gronovius met Zijn Kinderen*, 1775, watercolor. (Courtesy of the Museum de Lakenhal, Leiden)

This painting was completed a year before the Declaration of Independence was signed. Look closely, and the spinning globe next to Gronovius shows the New World with no United States as we know it now but instead a North America carved up among warring empires. We are going to take apart this world. In the following pages, we will open its cabinets and touch the specimens inside, unseal its jars to count the scales and fins, flip through the pages of its books and manuscripts, scrutinize its beasts immortalized in images, feel the bodily pains delivered by some of its creatures, and hear whispers of revolution from its animal protagonists. Animals creep along the margins of history and the fringes of documents, present but often undetected, influential yet unacknowledged. They linger beneath history's surface, sometimes materializing for just a moment—in a sidelong comment about wolves in colonial records, as the leather binding of a book, through a sloughed skin that made its way to a museum, or by staring back from a picture.

We will rove between the world of real-time encounters with live animals in the field, sea, and underground and the world brought back into the study, where naturalists like Gronovius wrestled with a cosmopolitan collection of specimens that even in death refused to remain quiet. In the process, we will follow an eighteenth-century geography: the geography of natural history. Although this book focuses on the early Americas and animal studies elsewhere in the vast British Empire, which stretched from the Atlantic to the Pacific, natural history was by definition and necessity multinational in the eighteenth century. Indeed, America itself was a created outcome of encounters with multiple "new" worlds (worlds only new, of course, to Europeans). We'll consider too, then, what animals and their study had to do with Big History themes like sketching out the maps of nations or charting the course of empire or the collisions between previously unacquainted cultures. In doing so, we'll hear about many of the human beings deliberately left out of such paintings.

This was a world of curious species, human and otherwise. Gronovius and all those lifeless animal bodies scattered around him in the portrait were considered such in the eighteenth century. The word "curiosity" has a convoluted and contradictory past. It has described rigorous scientific inquiry, yet also frivolous questioning. Darwin's virtue and Eve's vice. Ceaseless devotion—the term derives from the Latin *cura,* for care—and idle dilettantism. The key to unlocking secrets of nature, and the thing that killed the cat. Slippery in yet more ways, in the eighteenth century it could act as both subject and object in a sentence, denoting a state of mind and a state of being. One could be curious *about* things, and, at the same time, a thing, or an animal—or even a person—could also *be* a curiosity. In 1763, Nathaniel Bailey's *Universal Etymological English Dictionary* defined curiosity not only as "Inquisitiveness, a Desire of knowing" but "also a Rarity or curious Thing." *Curious Species* uses that slippage to open up an equivalent narrative space for animals. The goal is not to erase humans; it's to uncover a more-than-human dimension to the project of natural history—one that demands the same kind of historical attention lavished on human feats. Some animals were curious species in both senses of the term, as we will see in the case of inquiring beasts who generated their own taxonomies of nature, with consequences for the humans who studied them.[5]

Indeed, eighteenth-century people did not view animals as passive objects or as blank canvases onto which humans could enact their will. They saw animals as powerful actors, at times even as adversaries. Some framed their relations with them in the gritty idiom of politics. The famed French naturalist Georges-Louis Leclerc, the Comte de Buffon, proved one of the most influential observers of animals not only to his countrymen but across the British Atlantic world through popular English translations of his multivolume natural history. Although Buffon was a devout believer in human superiority, he claimed that man's "empire over the animals is not absolute. Many species elude his power, by the rapidity of their flight, by the swiftness of their course, by the obscurity of their retreats, by the element which they inhabit." Other animals, he observed, escaped human searching "by the minuteness of their bodies; and others, instead of acknowledging their sovereign, attack him with open hostility. He is likewise insulted with the stings of insects, and the poisonous bites of serpents; and he is often incommoded with impure and useless creatures, which seem to exist for no other good purpose but to form the shade between good and evil, and to make man feel how little, since his fall, he is respected."[6]

One might surmise from Buffon's tone that animal resistance made him sweat. Buffon's anxieties were typical. But paying attention to animals also reveals a brand of scientific inquiry quite different from those held in the popular imagination or typically studied by historians of this era: one forced to bask in what humans could not understand, reason out, or preserve. According to the standard narrative, a drive for total and totalizing knowledge of nature defined science in the Age of Enlightenment. Naturalists and natural philosophers attempted to master the nonhuman world by making it named, classified, stable, and thus subservient to the dominion of humans. In practice, however, naturalists were always coming up against the limits of their own knowledge, especially when it came to the canvas of willful, kinetic animals they studied. They doubted the capacity of their science to ever be all-knowing or all-seeing. That animals were a little bit hard to read was part of what made them curious.[7]

Until the past few decades, animals were largely written out of capital-h History, seen as too marginal to merit serious discussion as drivers of historical change. This book seeks to recover the capacities of animals that early naturalists sometimes called agency, sometimes volition, but most of all power: a formidable and frequently opposing force to human will and history. The knowledge of animals themselves and their impact on human knowledge-making existed as an inaccessible realm that couldn't be fully incorporated into natural history's body of facts. It sat there uneasily, as a kind of irritant, or spur, or parasite. Animals shaped the very project that studied them, setting the terms for what people could learn. If humans were at the wheel in the drama of the founding of modern science, there was often a mercurial raccoon riding shotgun, ready to pull the brakes or steer the whole thing off course.[8]

For all its focus on illumination, Enlightenment-era science found itself staring into the darkness.

In this book's early days, I envisioned it as purely historical, something I might corral within the strange and distant confines of the eighteenth century, like a lion caged in a courtly menagerie. If that era set practices in motion still in place among scientists today, it is also remarkable for its radical difference from modern biology and modern society. If its innovations leave a sticky residue on our political formations, our secular institutions, our naming of creatures and landscapes, our methods of study, or simply our inability to ever fully grasp the nature of animal life, many facets of that age are the proverbial foreign country; they did things differently there, merging the study of nature with the study of God as much more the rule than the exception, and braiding their natural historical findings with imperial expansion, political revolution, and the enslavement of fellow human beings. The term "presentism"—at best, histories that speak to contemporary issues, or at worst, an anachronistic collapsing of time that reads present-day worldviews and values back into earlier periods—can be a dirty word to some in the historical profession. Skeptics are right to note that we err when we generalize, when we lose the particularity and pastness of the past.[9]

The animals, however, had other plans for me. Over the course of research trips spanning more than twenty archives and three continents, I began to feel a pull not just from historical creatures in books and boxes but from those still out there in the flesh and the people who presently study them. Initially indulged in for "additional context," my field encounters and interviews soon seemed essential to foreground. And it was probably around the time I had my finger in the mouth of the world's rarest raccoon that I realized *I* wouldn't be getting out of this book, either, as one of the latest in a long line of curious species. Animals powerfully affected eighteenth-century natural history. They guide the work of modern biologists. Is it any wonder they would change my approach?

In form, *Curious Species* alternates between historical chapters about particular animals in the eighteenth century and narratives of my own interactions with the living inheritors, both human and nonhuman, of early natural history. That the animals were the ones to lead me here is, I think, significant. It shows that there's something stubbornly unresolved about human relations with animals from three centuries ago and that their history might bear on the environmental plight, a fully existential plight, plaguing the lion's share of the species discussed in this book. The present-day chapters herein, my accidental ethnographies, are narrative breaches in both senses of the word: a coming up for air from a submerged history to the surface of the present and a transgression in the traditional format of historical writing.

These breaches play several roles. One is a matter of historical argument and content. As we will see, eighteenth-century natural history generated an impasse about the knowability of animals that continues to shape our relationship with the descendants of these creatures today, requiring an interweaving of past and present if we wish to reckon with their uncertain futures. Another is a question of historical method. The lives of animals are hard enough to fathom when they are right in front of us. To access the fragmentary lives of bygone animals, we must turn to tools outside the usual parameters of historical archival and documentary evidence. Reenactment, embodied experience, literary devices, and engagement with contemporary scientific literature can offer multiple entries into a nonhuman animal's perspective and sensorium, however limited that picture might be. Animal history invites an interdisciplinary approach: ursine omnivory of sources and knowledge bases over koala-like specialization. Animals, in short, ask historians to write differently.[10]

A first-person perspective can also more fully encompass the *human* practices of the study of nature. One thing has held steady in some form since the eighteenth century: our dealings with animals are deeply personal. Enlightenment-era naturalists employed what we might now call subjective methods. Their accounts of animals were intimate, anecdotal, and emotional, even under controlled experimental conditions. Knowledge was situated and embodied, rooted in time and place. The perceived veracity of a naturalist's claims depended on his or her identity and subject position. Naturalists lived with, mourned over, touched, smelled, ate, dreamed about, and delivered the babies of their animal subjects to learn more about them. They reported their feeling-filled findings in scientific journals and other formal and distinguished venues. People could delight in not knowing, and many did quite unreasonable things in this self-described Age of Reason. "The modern idea of science as a purely rational, detached exercise was nowhere to be found in early empiricism," writes Meghan K. Roberts in her study of how French Enlightenment savants established their credibility through emotion, passion, and sentiment, especially by incorporating their families into their research agendas.[11]

As you will see, I have taken a page from them there. Indeed, we no longer have, if we ever did, the luxury of extracting ourselves from the matter. Over the past three centuries, human beings—though unequally so—have created the catastrophes of climate change and environmental degradation driving countless animals closer to extinction. The deep-sea bamboo coral on Gronovius's mantel, twisting like a banded sea snake and liable to glow in the dark when alive, is one of many species now imperiled by unregulated bottom trawling. And over the past fifty years alone, an estimated 70 percent of the world's sharks and rays have declined in abundance. If shark bellies can be seen as a metaphor for, but also the physical embodiment of, the hidden recesses of animal knowledge, of animal matter and mattering, what does it mean that we are now hemorrhaging such reservoirs at rates that lead many biologists to class our current era as a mass extinction in league with the one that took out non-avian dinosaurs some 66 million years ago? There is an urgency to our current situation that will not abide neutered prose.[12]

More than anything, we can never get away from the impossibility—in the eighteenth century or now, as naturalists or historians—of ever successfully effacing ourselves from our research in the name of objectivity. We cannot escape the animal foundations of human knowledge, not least because we ourselves are never not writing or working or thinking as animals.

This book, then, is not only a book about historical and contemporary natural history. Surely it is that: an account of the messy, serendipitous, embodied tradition we commonly call science. But this book is also about how research and writing get done more broadly, including the historian's craft. It's about the nature of knowing, how a new way of knowing was forged in the eighteenth century, and how animals were integral to that process. It splices the genres of history and natural history with the hope of troubling them both.

What do coral polyps, rattlesnakes, stingrays, and raccoons have in common? On the surface, not terribly much. They all swim, some more deftly than others. Each can birth live offspring in some form. All of their names contain the letter *r*. In the jargon of cladistics, modern biology's evolutionary-based classification system, all four descend from an ancient common ancestor, creating a so-called clade, an unbroken circle drawn around a large branch in the tree of life, this one labeled Animalia. Merriam-Webster's definition of the word "animal" rattles off a laundry list of traits—cells without cellulose, a need to consume proteins, the absence of photosynthesis, complex organization, "the capacity for spontaneous movement and rapid motor responses to stimulation"—a list that has only added several line items to the one drawn up by Aristotle in the fourth century BCE. Animal is a slippery category: impossibly various and, depending on the speaker, a negation, a big pie missing a human-shaped slice.[13]

By the end of the eighteenth century, naturalists classed all four of these as animals, too, though corals only belatedly so. *Curious Species* offers a deep dive into these four creaturely characters and their histories to consider some of the major problems plaguing the pursuit of scientific knowledge and animal nature in the eighteenth century and now. To add yet one more wrinkle to the shifting definition of curiosity: in the realm of the early sciences, curiosity could specifically refer to close attention to seemingly small, seemingly trifling things that through their particularities revealed larger truths, like the rush of blood glimpsed through a microscope or the peculiar glint of a butterfly's scales. Corals, rattlesnakes, fish, and raccoons were all especially intensely studied and debated as part of natural history. This deliberately mixed bag of curious creatures helps capture the diverse ways animals have shaped human knowledge. Rest assured, however—if these four animals are the main cast in the following pages, there are plenty of side characters, bestial Rosencrantzes and Guildensterns, waiting in the wings.[14]

Some of the beings most slippery to natural history, despite having been on the scientific agenda since at least the early modern period, have exerted a profound effect on knowledge about nature. They have forced naturalists to accept humanity's—even inquisitive and technologically sophisticated modern humanity's—limits to knowing. The creatures featured in this book exemplify various dimensions of these limits. Each section offers, on one hand, a chapter about how eighteenth-century naturalists grappled with a particular animal barrier and, on the other, how contemporary working naturalists continue to confront this now urgent (if long-developing) history of knowing and unknowing, especially as it relates to the high stakes of species loss in the twenty-first century.

These four sets of animals presented different puzzles to eighteenth-century natural history and reveal different facets of its practice. Corals made humans question: well, what even *is* an animal? What were the limits of categorization itself? And how could something like a brainless, diminutive polyp shape the fates of people? Rattlesnakes, in contrast, goaded investigators to ask: how might naturalists manage to study something that terrifies them? Would it be better *not* to understand a being's true nature, to instead embrace ignorance and fear? Fish demonstrated all that could go awry when one tried to contain nature in a book or archive. They spoke on one hand to the evanescence of life and on the other to the power of specimens to raise hell beyond the grave. (In 1765, naturalist David Skene grumbled to a peer: "I find my dead beasts more expensive than my living ones.") Finally, handsy raccoons, creatures both familiar and strange, raised questions about the thin line between human and nonhuman curiosity and the animal nature of knowing.[15]

Even as these four animal histories roam across the eighteenth century, the center of gravity of each section inches forward in time, leaving our historical terminus in the early 1800s. Each of these animals, however, also has an unfinished story. Biologists still contend with the outsized impact of coral and their blurry state of being, all while the same reefs that menaced early voyagers now suffer the arrows of bleaching, ocean acidification, and rising sea temperatures. People still propagate ignorance of rattlesnakes to aid their survival in the wake of climate change and targeted destruction. Some fish still resist the dried archive and artistic attempts to trap them on paper. A now critically endangered raccoon species is still unabashedly curious, urging us to contend with the multidimensional face of animal extinction and ecological collapse today. We will glimpse the historian in her habitat as well, likewise struggling to understand these animals through adventures dodging reefs by ship, tracking mother rattlesnakes in the fraught early days of a pandemic, attempting the lost art of fish flattening, and chancing a rendezvous with the golden-tailed raccoon.

Naturalists in the eighteenth century, moreover, sorted these animals in a hierarchy known as the Great Chain of Being, or *scala naturae*—an ordering of nature with vital importance to early natural history. Its character guides the structure of this book. Though the notion of a Chain of Being dates to antiquity, the eighteenth century saw the concept's zenith and most forceful application, particularly in the realm of natural history. The chain arranged the contents of the universe in a linear ladder, moving from lesser to greater complexity or "perfection" as one ascended. Inorganic materials formed the base, then plants formed the next rung, then animals, and then humans, who reigned over creation at the peak, as exemplified in illustrations accompanying the work of the Genevan naturalist Charles Bonnet (figure I.2). In some versions, angels stood above humans, and God ruled the chain from the top.[16]

Species today are understood to be dynamic and mutable. By contrast, most visions of the Chain of Being saw natural entities as static and fixed, plugged into their proper places along the chain, with infinitely fine gradations connecting the whole into a cohesive and awe-inspiring tapestry that might be intelligible in all its glory if only natural history could identify the missing links. Coral polyps and their freshwater relatives, the hydras (also called polyps), appeared to join animal and vegetable nature. Asbestos, a fibrous mineral, served as connective tissue between the plant and mineral worlds. Unions spiraled in every direction: amphibians marrying water and land, flying fish sea and air, bats mammals and birds. The Chain of Being was not just some concept in the clouds but a practical framework relevant to the workaday toil of naturalists on the ground while they sought to sketch out the underlying order of nature and marveled at the internal continuities of God's creation as they studied it in their own hands.

Indeed, it is largely because of the Chain of Being that animals achieved special status in the eighteenth-century study of nature. Even though many of the period's leading naturalists were primarily botanists, plants—whatever their beauty, whatever their imperial utility, whatever their effects on human bodies—did not abut humankind in the Chain of Being. The human mind could never be cleanly separated from its animal housing in the course of zoological research. Interacting with animals forced people to realize how animal

Figure I.2. A visualization of the Great Chain of Being, or *scala naturae*. This hierarchical diagram runs the gamut from crystals to man, whose half-clouded head reflects his awkward station as both a material animal and a cerebral, spiritual being. (From Charles Bonnet, *Oeuvres d'Histoire Naturelle et de Philosophie* [Neuchâtel, 1781], vol. 4, part 1, page 1; courtesy of the Clendening History of Medicine Library, University of Kansas Medical Center)

they were themselves and how woefully beyond human control the Chain of Being remained.

Although every vision of the Chain of Being bore its own peculiarities, one common theme was humanity's ambiguous position. For many naturalists, humans might have stood above the other animals, but they still partook of their being. In theory, the continuous, finely gradated nature of the chain did not permit drastic breaks or jumps, even though those on the far end of the spectrum, like Buffon, would make exceptions for the singularity of humanity. For others, the picture was much muddier and commingled, as perhaps captured best by the wonderfully complex run-on sentences written by naturalists as they struggled to commit the chain to words and define humanity's place in it. Take this flowing pronouncement of the English essayist Soame Jenyns: "The boundaries of those qualities, which form this chain of subordination, are so mixed, that where one ends, and the next begins, we are unable to discover. . . . Animal life rises from this low beginning in the shell-fish, through innumerable species of insects, fishes, birds, and beasts to the confines of reason, where, in the dog, the monkey, and chimpanzè, it unites so closely with the lowest degree of that quality in man, that they cannot easily be distinguished from each other." Because of the Chain of Being they constructed, humans were perpetually at war with themselves, struggling to reconcile the two sides of their nature. In seeing they were bedfellows of the animal world, naturalists had to thoroughly question their supposed domination over or separation from it.[17]

One chilling solution they devised to find peace with this humbling of humanity was to internally differentiate the human species. By applying similar hierarchies to human beings, naturalists could position people other than those arranging the chain nearer to animals as a buffer. Jenyns, for instance, continued his explication in racialized terms, writing of African peoples whom he likened to beasts: "From this lowest degree in the brutal Hottentot, reason, with the assistance of learning and science, advances, through the various stages of human understanding, which rise above each other, till in a Bacon or a Newton it attains the summit." By using people of color as collateral to shield European men from animals in this classification of life, natural history became a bubbling crucible of a nascent racial science.[18]

It is always worth remembering that the study of animals in the eighteenth century took place in the context of imperial land grabs and deep reliance on the slave trade. Knowledge of animals would not have flourished in the way that it did without human suffering. Natural history served as both beneficiary of and handmaiden to empire. Given the inseparability of natural history from the larger social projects and ills it either supported or relied on, animals both facilitated and thwarted those endeavors as well, and they did so in species-specific and historically particular ways. Thinking about animals does not obscure or divert attention away from the atrocities that women, people of color, and other groups faced. Quite the contrary: it can help bring those injustices to light. This is true not because these people were somehow closer to nature, as European naturalists so often asserted. Rather, it is because people on the margins of science and society—from women to children, from enslaved to free Africans, from Native Americans to Aboriginal Australians—were closer to knowledge production and were more essential than some historians have appreciated. They robustly theorized about nonhuman worlds and provided vital expertise to naturalists. On a material level, they hunted, captured, reared, maintained, trained, and preserved animals.

The past two decades of historical scholarship have seen a dramatic expansion in the human scope of who should count as a knowledge producer in the Enlightenment era, including people persecuted by natural history itself, viciously reduced into its objects and specimens of analysis. Focusing on historical animals forces us to think about process, which illuminates the real physical labor, not merely the far-off intellectual sorting, that people from all walks of life expended to make natural history work. Indeed, in some cases, we must resourcefully dip outside the eighteenth century to find the voices of people who have left fewer written traces from the period—not least due to anti-literacy laws that forbade reading and writing among the enslaved in some regions.[19]

This book's four parts work up and through the Chain of Being—moving from corals to rattlesnakes to fish and finally to humanoid raccoons—not to endorse that hierarchy but to excavate and challenge it. This organization immerses and roots us in the worldview of Enlightenment thinkers while also critiquing it from within, finding instabilities lurking inside the scheme. By the end, readers will see how the animal seeps into the human. All humans.

If the Chain of Being or Gronovius's cabinet seems a stark departure from the twenty-first century, there are significant ways in which modern biology inherits this era. Not even the Chain of Being is completely defunct. On occasion, I could see it quietly at work as I interacted with practicing scientists while researching this book. At the start of one collection visit, a biologist in his office handed some historical fish specimens to me and my husband, William, grabbed a few more for himself, and with little warning called over his shoulder for us to follow him to the museum's basement storehouse. Never had I had such an experience at a traditional archive of paper documents. We began running—literally running—in a desperate attempt not to lose sight of this man or lose hold of our dried fish, down flights of stairs, through back corridors, past shipping crates and forklifts and a giant squid putrefying in what must have been one of the longest single-occupant specimen jars in existence. A flounder went flying from the man's hands—pick it up, dust it off, run along. For readers who have played Nintendo's *Legend of Zelda: Ocarina of Time,* it felt like trying to keep up with Dampé, an impossibly fast ghost one must chase through the crypts to secure a valuable item of gameplay.

En route to see the historical fish specimens on our agenda, the biologist stopped as suddenly as he had started. He said he had a secret to share, and he apparently deemed us trustworthy. We took a detour into the deep catacombs of mammal storage. In the dim and dusty light, he cranked an opening into one of many towering steel cabinets and pointed to a bottle sealed with a red lid made of stretched pig's bladder. Here vagueness will assure anonymity. Inside the jar was a small mammal submerged in alcohol, eyes closed and legs folded. It had been in that bottle for a good three centuries, likely originating from Albertus Seba, a Dutch naturalist with a series of famed collections that have since undergone several successive and hard-to-trace dispersals across European museums. We would later, with little ceremony, view fish and reptile specimens just as old, with the same red lids, likewise suspended in fluid—wine, brandy, and rum being the preferred preservatives of eighteenth-century naturalists. But our host wasn't supposed to show us this treasure purely because it was a mammal, the unstated assumption being that some thought it deserved a more dignified medium of rest. There still reigns an implicit belief that some animals are higher, some lower.[20]

This specimen's verboten nature wasn't a one-off. Take the natural history collection of King Adolf Fredrik of Sweden. In the eighteenth century it housed a miscarriage from Queen Lovisa Ulrika, his wife and fellow naturalist-collector, stored with other human and nonhuman fetal specimens behind a curtain to conceal them from the unprepared eyes of ladies. Though the collection holds no humans today, it is now partially digitized, and the website warns visitors of the remaining fluid-preserved mammal embryos and primates they are about to witness: "Please, be advised that the collections contain animals of many different groups. . . . Some may find these images macabre or repulsive. As scientists and curators of these specimens, we maintain them with all due respect. They generally represent a practise in natural history cabinets that is no longer pursued." Wet coral specimens generally receive no such content advisory. Despite attempts to distance this past, modern viewer sentiment toward animals lingers in the shadow of the Chain of Being.[21]

Surely it remains at work in our own species as well. Humans converted other humans into specimens during the eighteenth century, especially, though not exclusively, if they were deemed monstrous or deviant or of a curious shade of skin. The snail's pace at which the bones of Indigenous and enslaved people have exited museum collections in recent decades, the absence as of this writing of any explicit federal legislative framework for the repatriation of African American remains or protection of burial sites, and in some cases the continued use of historical human bodies in teaching and display without so much as a shred of something resembling consent offer natural historical manifestations of how the dignity humans show others of their species continues to run along a spectrum congruent with the hierarchies devised by early modern science.[22]

Beyond its Chain of Being, the science of the eighteenth century shapes the practice of modern biology in more fundamental, even routine, ways. Pull up the Wikipedia page for many of your favorite animals—domestic cats if you're like me, boa constrictors if you're brave, human beings if you resemble many of our kind—and in the right-hand column you will see a two-word Latin name, such as *Homo sapiens,* followed by the notation Linnaeus, 1758. Thousands of the official species names recognized by scientists all over the world today were willed into existence by the project of Enlightenment classification, most clearly embodied in the work of the Swedish naturalist Carl Linnaeus. Linnaeus's *Systema Naturae,* "The System of Nature," presented a sweeping arrangement of the known natural world and a practical guide for ordering animals, plants, and minerals. First published in 1735, its monumental tenth edition of 1758 employed a "binomial" or two-word system of naming organisms, as opposed to the previous practice of longer Latin descriptive phrases. Many species of corals, snakes, fish, and raccoons were first formally named and described in some edition of Linnaeus's work.

Although Linnaeus was not technically the first to employ binomials, the tenth edition of the *Systema Naturae,* along with his botanical work *Species Plantarum* in 1753, made the use of a Latin genus name (such as *Homo*) followed by a subordinate Latin species name (such as *sapiens*) the gold standard for scientists thereafter. In cases where Linnaeus or other naturalists named a species only for it to later be shuffled into a new genus, the notation will now appear in parentheses, such as (Linnaeus, 1758). And for the sundry species not known to Linnaeus or those that were formally deemed separate species by other naturalists after him, one might instead find a notation such as Cuvier, 1817, as for *Parastichopus regalis,* the royal sea cucumber, or Helgen, 2013, for *Bassaricyon neblina,* the olinguito, a small carnivorous relative of raccoons recently identified as a new species based on genetic testing performed on museum specimens, collections being the new proverbial field of discovery. But the general practice has stuck. Every working biologist or taxonomist alive today knows the name of this eccentric eighteenth-century Swede.[23]

None of this is to naturalize Linnaeus as a rupture with the past or uphold him as the germ of the one true science. Despite his influence, Linnaeus had major detractors in his day, such as Buffon, major predecessors such as John Ray, and major collaborators, informants, and sources he relied on who are rarely part of the scientific lingua franca, as with Kwasi, a former slave turned man of letters in Dutch Surinam who discovered the bitter medicinal shrub amargo. (Linnaeus christened the plant *Quassia amara* in acknowledgment.) That designation—Linnaeus, 1758—brings a finality to the ordering of the world that wasn't faithful to either the complexity of nature or that of the people who studied animals beyond Linnaeus. Indeed, this book seeks to open up the standard picture of early natural history as the pursuit of a narrow cohort of elite European men by also locating enslaved taxidermists, Native interlocutors, female chefs, and even child collectors in the story. But those two words—Linnaeus, 1758—give one specific example, among many explored in the following pages, of how we continue to operate in the categories and frameworks of Enlightenment-era science.[24]

Something changed in the European study of animals during the long eighteenth century, especially near its midpoint. During the earlier Renaissance, an era of wonders and marvels, the scope of the world known to Europeans unfurled like the maps they would draw of it. With long-distance seafaring voyages came an expanded horizon of animal nature and an influx of information about strange creatures, useful medicinal plants, and anthropoid monsters. Naturalistic description, depiction, and nomenclature mingled with emblem, story, and the fantastic in this stream of animal study. A woodcut illustration made by Albrecht Dürer in that age is perhaps the most famous animal image of all time: a rhinoceros who seems part pachyderm, part dragon, part armored knight, part unicorn at the shoulder (figure I.3). This beast, named Ganda in its homeland of northern India, was a real, breathing animal and individual subject, as evinced by that eye fixed off-canvas in stern yet weary resolution. Ganda made a grueling maritime journey to Lisbon in 1515, served as a living gift to a king, and was set to do the same for a pope before dying by shipwreck on the journey to Rome and taking on a new life forever after in images. Yet legend and symbol stick to Ganda's carapace. Little resembling the colossus in Dürer's print remains in Gronovius's cabinet.[25]

Eighteenth-century natural history was never quite the "objective" enterprise as characterized by an older guard of historians. And we will soon see that symbology and story still figured centrally in the study of animals in this era, perhaps best exemplified by the case of the rattlesnake. Yet animals did, in some significant sense, get estranged under Enlightenment-era classification, collection, and the strong arm of system. They became potentially unknow-

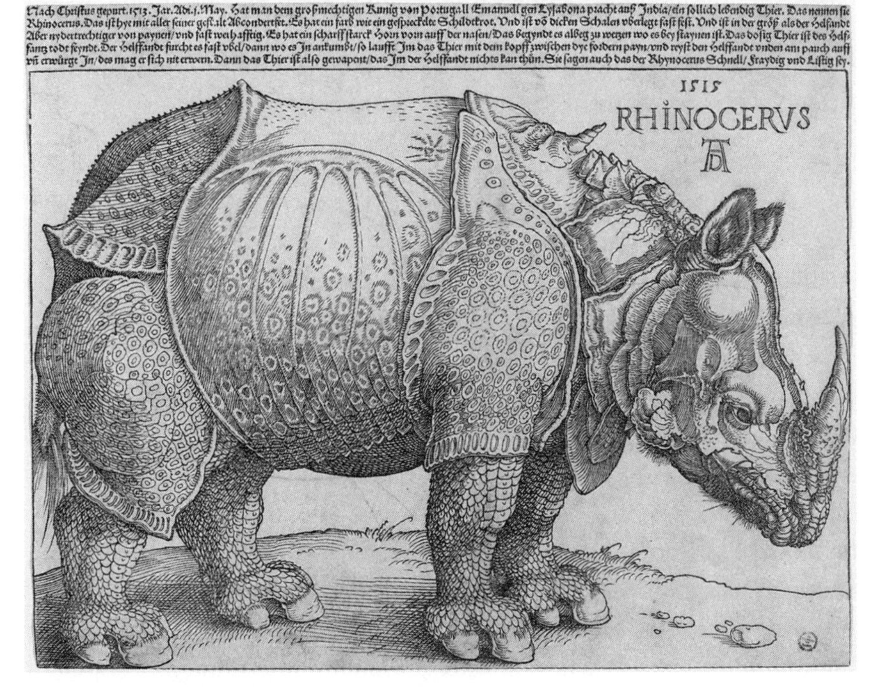

Figure I.3. Albrecht Dürer, *Rhinocerus*, 1515, woodcut on laid paper. (National Gallery of Art, Washington, D.C., Rosenwald Collection, 1964.8.697)

able because they weren't allowed to speak to us in the same way as they had before. They became objects of natural history—a project that was less anthropocentric on one level and also regarded humans as animals, but a project that was nevertheless a reduction all its own, an attempt at disenchantment. That process, in turn, created an impasse about what animals are and what they might truly be capable of—and how humans should even begin to answer those questions—that has never been resolved.[26]

The twenty-first century inherits that impasse. We must look animals in the eye and realize how little we know of them, not least those on the verge of disappearing for good. No perfect severing or ordering of the universe followed the invention of Linnaean taxonomy or the formation of modern museum collections or the rationalization of the sciences. Naturalists are ever subject to the terms of nature, caught in the double bind faced by their predecessors. We are damned with animals, damned without them.

■ ■ ■

This book studies the activities of a discipline, but in so doing it also aims to chip away at the strictures of discipline. Today's scholarly silos of knowledge would be unintelligible to eighteenth-century people—not only those between one science and the next, but also the firmer walls cordoning off art from science, literature from science, and religion from science. The same people who studied corals likewise studied fish and elephants and medicine and ancient druids and poetry and God. We will meet naturalists in the following pages who took on or combined any number of vocations and identities: noblemen and ruffians, travelers and homebodies, planters and plant collectors, printers and pastors, professors and jurists, physicians and philosophers. My present-day interlocutors mirror the capaciousness of naturalists past. I spoke with field biologists, taxonomists, museum curators, ship captains, conservationists, artists, fishers, fish mongers, and tour guides with intimate local knowledge. Although we will find much to critique about the imperial modernity forged by Enlightenment-era thinkers, this book is a bid for their gluttonous approach to inquiry, suggesting what might be gained by putting separate realms of knowledge in conversation.

Curious Species is an experiment in form for thinking through how we might strategically be undisciplined like the most errant of animals. In particular, it borrows tools from a tradition of inquiry that, unlike professionalized history, never needed a reminder to remember nonhumans: nature and science writing. Early New World travel narratives tended to note as much about the fauna and flora of freshly encountered lands as they did their peoples. This would remain a robust Anglophone literary tradition, from the likes of Henry David Thoreau at Walden Pond in the nineteenth century to prophets of nature study such as John Burroughs and John Muir at the turn of the twentieth, to today's nature and science writers, from Robin Wall Kimmerer to Helen Macdonald and David Quammen. As a form of creative nonfiction, the best nature and science writing often entertains perspectives that can never be experienced by mere humans. It relishes radical reconsideration of the possibilities of the written word by imagining the aerial view of an albatross or the knowing grasp of an octopus arm.[27]

Take, for example, Verlyn Klinkenborg's rigorously researched *Timothy; or, Notes of an Abject Reptile* (2006). Written from the vantage of a real tortoise that once roamed the garden of Gilbert White, an eighteenth-century English parson-naturalist, the book consists entirely of short, choppy sentences. Many of them fragments. They read as bursts of measured animal thought, and, like cud, should be chewed, mulled, and digested slowly. Each era's brand of nature and science writing can be historicized in great detail. Like natural history proper, their American and European manifestations were forged through colonialism—and I remain a settler-observer. But rapt attention to a buzzing, biting, slithering nonhuman world underwrites it all. Honoring the specificities of historical archival material need not be incompatible with writing in a way that caters to the temporal and subjective experience of other animals.[28]

Historians are like pigs, if they'll forgive me for saying so. This is, from me, a compliment. They are intelligent and formidable. Their memories stick. They travel in packs and zealously defend their young, their protégés. When rooting around for things beneath the surface, they leave no stone, or piece of evidence, unturned.

Nature and science writers, on the other hand, are more like bees. They generate buzz. Some are specialists, some are generalists. They, too, form a community. They gather nectar from many sources and transmute it into something that is more than the sum of its parts.

This book is a hybrid monster, born of two genealogies. I hear it rumbling now. Can we make a pig fly?

CORALS

We are lost in astonishment while we contemplate the nature of Life: the deeper our inquiries go, the farther does the object seem to recede from us.

—*John Graham Dalyell, "Physiological Reflections on the Natural History of Animals and Vegetables"*

CHAPTER ONE

A Different Species of Resistance

AS JAMES COOK MADE HIS WAY up the eastern coast of Australia during the first of his three famous voyages, he almost lost his ship to the largest animal community on the planet. In June 1770, the HMB *Endeavour* smashed into the massive, recalcitrant Great Barrier Reef: a maze of underwater matter made by tiny animals called coral polyps. To the minds of naturalists in the late eighteenth century, these brainless polyps were the nadir of the animal kingdom. And yet, they unwittingly steered the fates of men.

London's esteemed Royal Society had sent Cook's expedition to join a multinational effort to calculate the distance between the earth and the sun by observing the transit of Venus. But the Pacific was also a theater of imperial jousting. Then still a lieutenant but soon to be promoted to captain, Cook had received secret instructions from the British Admiralty to search for the rumored *Terra Australis Incognita,* a hypothesized sprawling continent in the southern hemisphere that, had it existed, would dwarf Australia in magnitude. As shown by his skirmish with the great barrier of the reef, a structure unfathomed by European science, Cook would have to contend with both human and nonhuman communities in that mission. And yet, when examining the ship's damage once safely on shore, Cook's crew found that although the hole "was cut away as if it had been done by the hands of Man with a blunt edge tool," some "Coral rock" had propitiously lodged in one of the vessel's lacerations, preventing a far more catastrophic leak that could have spelled the end for the expedition. Voyage artist Sydney Parkinson, who would die of dysentery only months later, thus concluded that the same coral "that endangered us, yielded us the principal means of our redemption."[1]

Corals obstruct. They and other animals made the discipline of natural history possible, facilitating understandings of the natural order in innumerable ways. But they also made this project fundamentally fraught. In Cook's case, lowly corals impeded the work of science and empire, threatening to erase both his men and their discoveries, while also saving the day with a deus ex machina: they maimed the ship and plugged the hole. More broadly, polyps and the reefs they constructed aided natural history efforts by becoming animate arbiters of the animal-vegetable boundary, igniting discussions and experiments that probed the essence of animal life and the wondrous creative potential of the natural world. However, their slippery nature also threw natural history into its own forms of chaos by endlessly blurring categories of being and bending the standard rules of existence.

At the time of Cook's voyage, early taxonomists had only recently shunted corals into the animal kingdom. In fact, Cook's clash with the reef culminated a longer eighteenth-century quest to study and confront the power of these organisms, along with the nature of other liminal beings. Corals, anemones, sponges, jellyfish, tube worms, Venus flytraps, and other vegetative animals and animalistic plants muddied life's tidy kingdoms. Some, like corals, seemed chimeras of animal, plant, and stone (figure 1.1). Though apparently only "half-animated," in the words of botanist James Edward Smith, all remained moving targets when studied, being creatures that crept and ruptured existing systems of classification in the process. Collectively, many were termed *zoophytes*—literally "animal-plants"—or in the case of reef-building corals, *lithophytes,* or "stony plants." A number of naturalists embraced these ambiguous terms to encompass the motley group of beings on trial for assailing the very nature of nature's categories.[2]

Even at the height of scholarship and learning that characterized the Age of Enlightenment, then, people struggled to answer a basic question: what is an animal?

This gray area of organisms provoked tortuous philosophical debates and increasingly absurd laboratory trials seeking to respond to this question. The answer is still far from straightforward. Today's biologists, like their precursors in the mid-to-late eighteenth century, also identify coral polyps as animals—those small, soft, daisy-shaped creatures that constitute coral "colonies" and

Figure 1.1. A dried coral specimen, absent its soft polyps. The skeletal remnant figured here would have been made of aragonite, a form of calcium carbonate (limestone). Holes across the surface represent corallites, the housings of individual coral polyps. Many early naturalists classified corals as plants due to the branching form of specimens like this, along with the floral appearance of polyps. (From John Ellis and Daniel Solander, *The Natural History of Many Curious and Uncommon Zoophytes* [London, 1786]; Biodiversity Heritage Library)

that, in the case of scleractinians, the so-called stony or hard corals, erect limestone labyrinths like the one that snared Cook's ship. But the promiscuous mixture inherent in the notion of zoophytes also finds some basis in what we (think we) know now of their nature. Coral polyps are themselves miniature ecosystems. The tissue of many species harbor plant-like algae, commonly called zooxanthellae, that provide color and nutrition through their ability to photosynthesize, giving their animal hosts a boost in building reefs. When stressed by changing environmental conditions, coral polyps eject these symbiotic bedfellows and turn bone white, creating the infamous spectacle of coral bleaching—the physical and visual manifestation of a ruptured nonhuman partnership. Corals, then, are seeming animals, containing seeming plants, secreting seeming stone (figure 1.2).

In the eighteenth century, the peculiar properties of these creatures drove inquiries into the category of the animal writ large, though the unstable nature of corals just as often thwarted such investigations. Perhaps the only constant about corals is their tendency for transfiguration, their capacity for contradiction. Their flesh is indeterminate, always both being and becoming. Polyps are discrete organisms, yet they are physically chained together with other polyps by live tissue known as a coenosarc. Some species can alternate between free-floating individual animal phases and aggregated communal states in response to environmental stress. Their social state changes throughout the course of their lives, as well: coral larvae are mobile and sentient swimmers who can sense the characteristic crackling of fish and crustaceans on their journeys to find reefs, where, buoyed by ocean currents, they arrive and settle down into stationary communities. An individual polyp might be male, female, or both, with fertilization occurring either inside its body or externally in the surrounding waters—that is, aside from those coral polyps who instead decide to reproduce asexually, even if they were themselves the product of sexual union. To replicate, such a polyp will bud a potentially infinite series of clones from the side of its body like a grotesque sci-fi magic trick, rendering these animals functionally immortal, something many early modern naturalists thought possible only of rational human souls. Corals are soft, minute, goo-like animals, yet they achieve the alchemy of biomineralization by forming reefs, what biologists call "the most important bioconstruction of the world."[3]

The ever unsteady, one-but-many nature of coral communities shaped how people studied and interpreted these animals throughout the eighteenth century, even if naturalists had only inklings of the full scale of biological

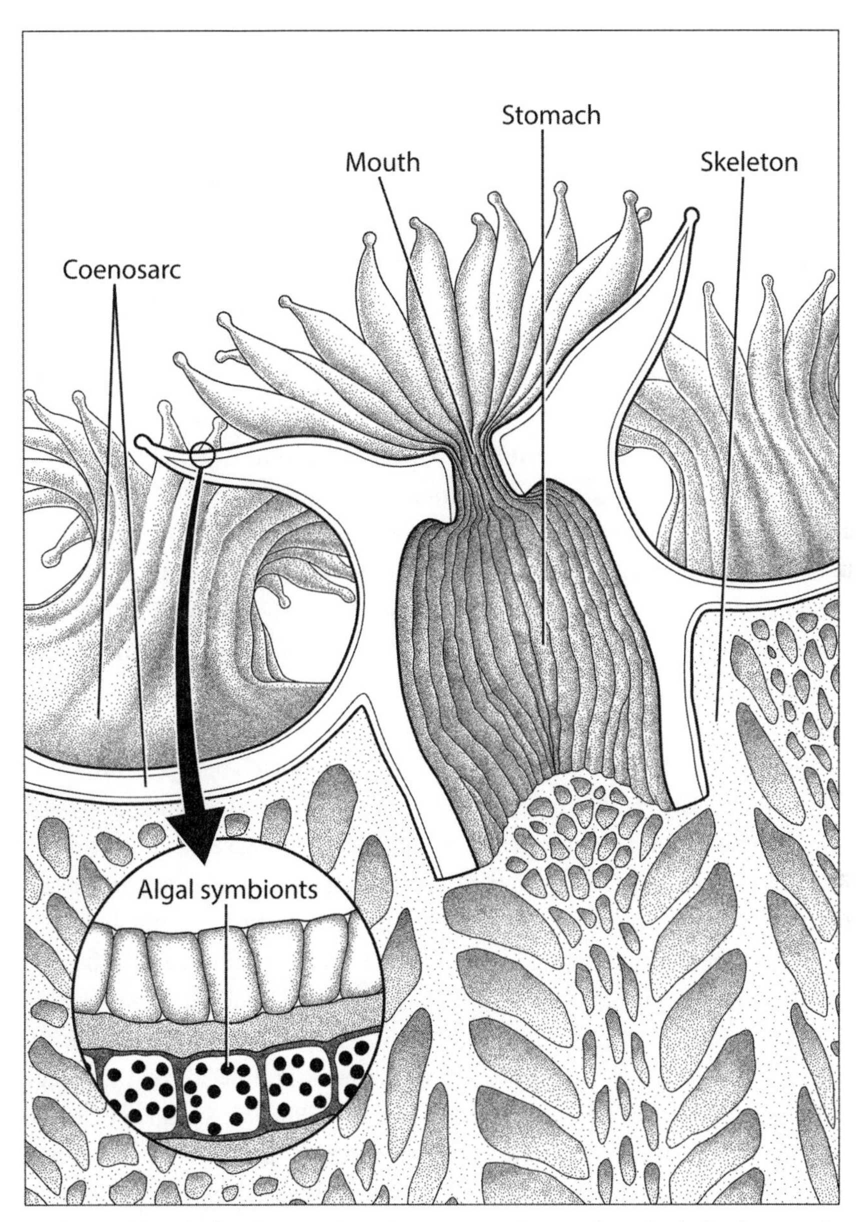

Figure 1.2. Diagram of a scleractinian coral community, showing the individual polyp animals, the limestone skeleton they secrete to produce reefs, algal symbionts (also called zooxanthellae), and a connective tissue known as the coenosarc, which physically joins individual polyps within a community. Each animal sits in a skeletal cup called a corallite. (Artwork by Emily S. Damstra)

mechanisms at work. Human projections about corals—what they thought these animals might have been doing, or what they might have meant for the natural order—always operated alongside actual corals, doing actual things in the sea. In the words of the political philosopher Thomas Hobbes, natural history was "the History of such Facts, or Effects of Nature, as have no Dependance on Mans *Will*." The shapeshifting flesh of corals inspired debates over animal will and animal power, given that delicate polyps produced stone monuments composed of their ancestors with the heft to bring down ships. Corals drew attention to the ultimate unknowability of animals, seemingly at odds with human efforts underway to comprehensively catalogue and describe nature.[4]

It was precisely these creatures' lowly status within a hierarchical worldview that made their actions profoundly unsettling to various eighteenth-century thinkers, be they naturalists, physicians, or clergy. Cnidarians—an ancient phylum of soft, radial beings with stinging tentacles that includes corals, freshwater hydras, and sea anemones—formed the group of animals stationed lowest on many versions of the Great Chain of Being. Some naturalists and philosophers conceived of them as a link between or mixture of animal and plant worlds. Corals were the limit case of animal nature, bringing the stakes of that category into relief. The seeming mismatch between minuscule polyp and massive reef prompted anxieties and debates surrounding the volition, handiwork, and consequences of animals. Naturalists such as William Stukeley, whom we will meet in the following pages, found it terrifying to ask: how could something so small make something so massive? Could the lowest animal on the Chain of Being have a will that restricted the wishes of people? Corals created things at odds with their own nature, yielding an uneasy gap between animal cause and effect. The barely perceptible polyp constructed the firm fortresses of reefs, the airy quills of sea pens, and brain coral's cerebral twists. Corals were of the sea, but they made land. Both biological and geological, they built worlds underwater and sometimes thrust above—from the jagged peaks of the Dolomites to some of the islands where colonialism took root.

The Irish-born naturalist John Ellis, a well-to-do fellow of the Royal Society and eventual imperial agent for several New World colonies, emerged as the star British voice on the animal nature of zoophytes. In 1755, he published his influential *Essay Towards a Natural History of the Corallines, and Other Marine Productions of the Like Kind*. In a section on sponges, another animal that doesn't behave like a typical animal, he described "a different Species of Resistance, from that which a Substance, not animated, could be supposed to do." Understanding this species of resistance—what seemed the origins of animal animation, will, and power—largely became the point of zoophyte studies. At the same time, however, the volition of corals and other zoophytes vexed their study *as* animals: their motion and sentience proved difficult to observe, preserve, and stabilize, leading many humans to class corals as stones, as plants, or as some monstrous mixture instead. The resistance of these beings profoundly shaped eighteenth-century thinkers, who came to recognize the capacity of such small animals to inhabit worlds of their own and hamper human designs in the process. They likewise realized that small-scale dynamics among "colonial" organisms, a nonobvious term *produced* by this colonial moment in history, might have import for large-scale human political configurations and perhaps the nature of God himself.[5]

The very foundations of the world were shifting in the eighteenth century. Philosophical, political, and social revolution loomed at every turn. Corals, modest as they look, had a role to play in the upending of an old order. As for natural history, over time, the unsearchable nature of corals and other zoophytes forced many naturalists to abandon their pursuit of rigid hierarchy, sowing the seeds for a celebration of animal power.

THE ANIMAL WITHOUT A FACE

Zoophytes confounded those who studied them in the eighteenth century. Language broke down when trying to capture them in words or images. Animals seemed to bud and stem, roots shooting and flowers blooming, while New World plants snapped flies in pink maws and soft bodies turned suddenly to stone. Although scholars tend to stress the early misidentification of corals as plants or minerals, that sells their forms short. These animals conjured far more than botanical or geological visions to natural historians: they were sea fans and sea pens and dead man's hands, memorialized in name as stars, church organs, insects, reptiles, brains, and even racialized bodies. Descriptions of zoophytes grasped widely for analogies that might make the alien natures of such beings concrete and intelligible. Such ambiguity would not abate with time in any simple trajectory. Charles Darwin, on the heels of the Enlightenment, mused that the "tree of life should perhaps be called the coral of life, base of branches dead; so that passages cannot be seen," given that the reef structures that polyps build—the congealed, rocky history of their ancestors—adhere to the latest generation, like the obscured cul-de-sacs and extinctions of evolutionary history. Horst Bredekamp has argued that Darwin based his "coral of life" concept on a specimen he collected in Patagonia in 1834 that Darwin believed, quite understandably, to be coral. But this too was confusion: the specimen turned out to be red algae.[6]

People had been acquainted with zoophytes in some form since antiquity. Aristotle's accounts of them remained a touchstone for eighteenth-century naturalists, as did his working definition of animals as creatures capable of sensation, digestion, reproduction, motion, and production of vital fluids like blood. Corals, specifically, were already a fascination of naturalistic investigation well before the Enlightenment. During the Renaissance, the crimson branches of the Mediterranean soft coral *Corallium rubrum,* crafted in dark caverns by ethereal white polyps, figured conspicuously in the collections of Wunderkammern, though with hardly a hint of their animal makers. Coral's classical and allegorical associations would have resonated with visitors to these cabinets of wonders. In Ovid's telling, coral originated when the hero Perseus placed the slain Medusa's severed head on a nest of beached seaweed, which hardened at the touch of the ferocious Gorgon—fitting with long-standing notions, dating back to Roman naturalist Pliny the Elder and earlier, that coral was a soft submarine plant that turned rigid and rocky once exposed to air. (Mythic overtones persist in coral nomenclature: today, many soft corals, such as sea whips, are called gorgonians.) Renaissance red coral could be wrought into jewelry, elaborate sculptures, devotional objects, protective amulets for children, and teething toys, becoming a precious material that embodied the playful artistry of nature and harbored curative properties. Physicians and apothecaries incorporated coral in a number of remedies. Theophrastus von Hohenheim, known as Paracelsus, recommended bright red coral to ward off melancholy and ghosts and to quicken the imagination. Even while its animal origins remained unknown, coral proved a potent and multifarious material.[7]

Corals commanded the attention of people beyond Europe as well. Red coral was prized both in West Africa and among enslaved African communities in the New World, so much so that coral, along with other would-be natural curiosities like cowrie shells, served as currency for purchasing people along African coastlines. Observers aboard Cook's *Endeavour* also noted how Indigenous groups throughout the Pacific used corals as technologies. Joseph Banks, the official voyage naturalist, described how Pacific islanders fashioned instruments such as fishing hooks using "files of Coral," which he noted "work in a manner surprizing to any one who does not know how sharp Corals are." (No one aboard the *Endeavour* would doubt coral's sharpness after the ship foundered on the Great Barrier Reef the following year.) Banks noted disdainfully that these communities built entire buildings, boats, and pavement with nothing but "an axe of Stone in the shape of an adze, a chisel or gouge made of a human bone, a file or rasp of Coral, skin of Sting rays, and coral sand to polish with," though the ray skin and coral, Banks admitted, made all these productions "very smooth and neat." His backhanded compliment showed that Pacific peoples quite smartly realized the tooling potential in coral's rocky texture and adapted it to their own ends.[8]

Given these manifold meanings of coral, its transformation into a cause célèbre for determining the bounds of animalhood in Euro-American natural history was neither natural nor inevitable. In fact, the squishy polyps themselves were largely seen as base and unworthy of contemplation before the middle of the eighteenth century. Jacques-François Dicquemare, a French astronomer and naturalist, spoke to the former disrepute of zoophytes in his study of sea anemones, close relatives of coral that are essentially blown-up polyps in form and can likewise regenerate upon cutting. Dicquemare noted in letters he penned to the Royal Society in the 1770s that these and similar creatures were "deemed hardly worthy our attention" in prior years, seeming "intended to be trodden under the foot and only taken notice of by chance." (Dicquemare also detailed how he repeatedly attacked sea anemones with scissors to monitor their regrowth, listing goals for his research program as varied as proving God's greatness through the meagerest of animals and divining anemone palatability by boiling twenty to feed to his trusty co-experimenter, the cat.) Even into the 1770s, Oliver Goldsmith, a poet, playwright, and novelist who authored an influential multivolume synthesis of natural history inspired by and adapted from the work of Buffon, continued to describe zoophytes like polyps as "nauseous and despicable creatures," which "excite our curiosity chiefly by their imperfections," even though they had the power to "totally overthrow" the cherished systems of naturalists. Dicquemare saw "an advantage reserved for" the eighteenth century: naturalists could turn their eyes anew to "introduce a new set of beings" as answers to pressing questions about animal nature and the generation of life—even if those creatures seemed squalid, foul, or worthy to be squashed, at first glance.[9]

Several developments near the midpoint of the eighteenth century led Europeans to reassess the nature of coral. The rise of Linnaean taxonomy devised more rigid and formal classification systems. The spread of European colonialism, especially on islands and coastlines, precipitated maritime coral encounters. But beyond these wider forces, interest in zoophytes began to take a concerted animal turn thanks to the day-to-day interactions between one curious animal and one curious man. In 1740, a Genevan naturalist named Abraham Trembley, while working as a tutor for two young children of a noble house in The Hague, conducted experiments on the regeneration of freshwater polyps and subsequently conveyed his findings to the French naturalist René-Antoine Ferchault de Réaumur. Also known as hydras, these small yet sturdy animals that can fit in a droplet are the freshwater kin of saltwater coral polyps, though they do not build reefs. The history of how corals came to be classed as animals in the eighteenth century is unintelligible without first considering their freshwater doppelgänger, which could be more readily studied and helped set zoological investigations of corals in motion.

Trembley was not the first to encounter such a being. Antonie van Leeuwenhoek, the secretive Dutch textile merchant whose handheld microscopes led him to discover no less than bacteria and sperm, observed and had drawings made of freshwater hydras in 1702. But polyps didn't make much of a splash until news of Trembley's experiments spread. Trembley found the sprightly beings, which resemble a thin rod topped with lacey tentacles arranged around the creature's mouth, in some ditch water by mistake (figure 1.3). And he did what any reasonable naturalist would do with such a find: he cut his specimen with a knife from every angle imaginable. To Trembley's amazement, and no doubt the children's, the severed pieces regenerated, without any copulation, into whole organisms that soon began to inch across his glassware like worms. In one experiment, by successively cutting, he brought forth fifty new animals from a single hydra before calling it a day. Trembley's research agenda converted his laboratory into a precarious clutter of ever multiplying glass jars to house the newly conjured animals.

According to Dutch botanist Jan Frederik Gronovius, many fellows of the Royal Society did not believe news of the discovery until they received shipments of the animal in the flesh and enacted their own violences on it, and, "after having well examined this Creature found the Prodigy of increasing it self in that Wonderful Manner very True." But if this was an animal, it was a bizarre one. Hydras, as with coral polyps, can reproduce sexually with eggs, but they often multiply asexually via plant-like budding by releasing a copy of themselves from their body, or through trauma such as human slicing. Despite a polyp's vegetable appearance and propagation, Trembley identified it as

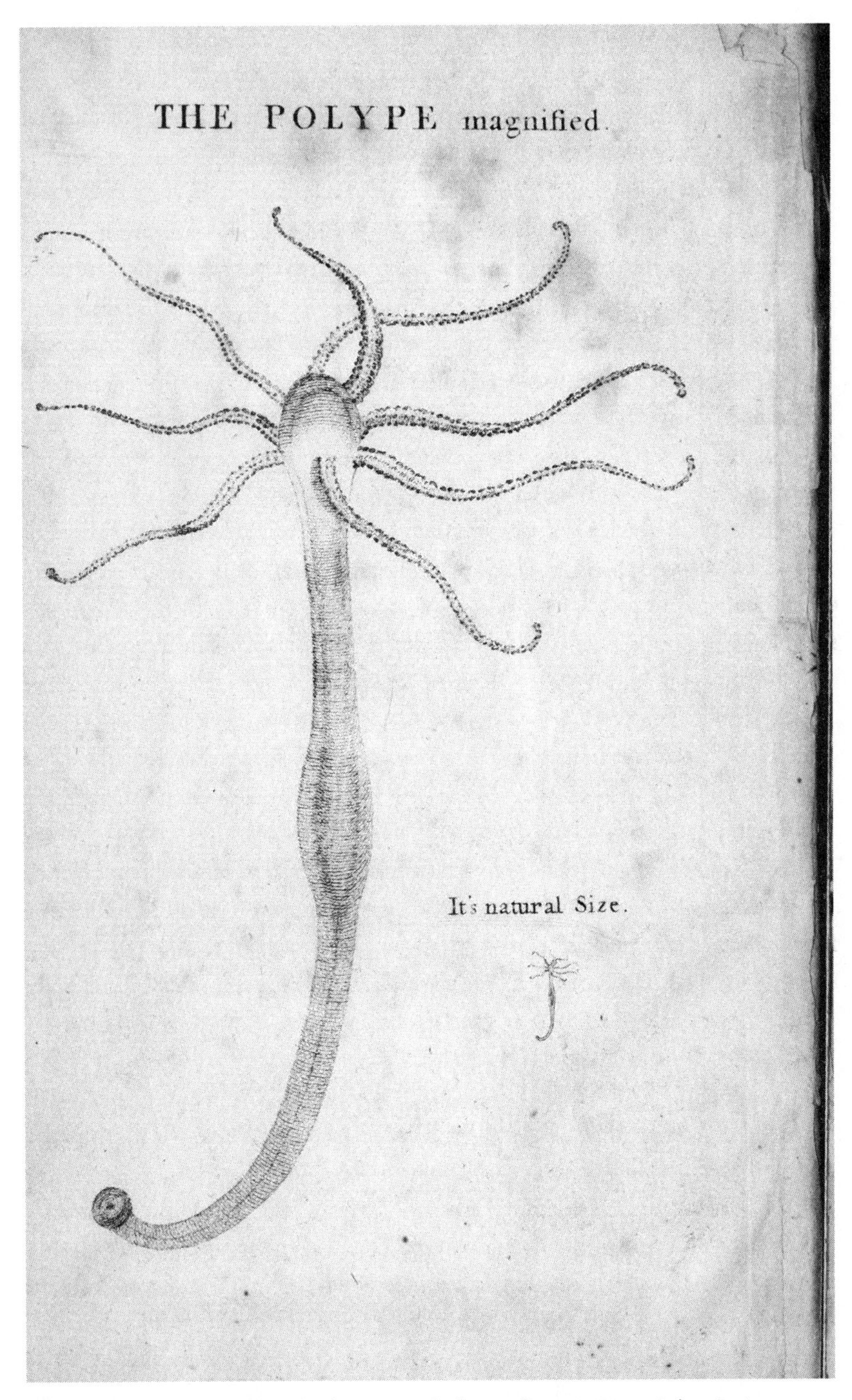

Figure 1.3. A freshwater polyp, also known as a hydra, as shown in Henry Baker, *An Attempt Towards a Natural History of the Polype* (London, 1743). (Biodiversity Heritage Library)

an animal based on its spontaneous motion, predation, and digestion, harkening back to Aristotle's categories. Its crown of soft threads acted as "so many Snares" to catch prey, in Trembley's telling. The creature was essentially "one great stomach," in the later words of William Frederic Martyn. When Trembley flipped the animals inside out, "as one may turn a Stocking," the creatures remained fully operational, "as if they had never been turned." Trembley's findings would motivate his cousin Charles Bonnet to chop up earthworms and the Italian naturalist Lazzaro Spallanzani to do the same to snails and salamanders, showing the regeneration of parts as complex as limbs and heads and brains. But a polyp could offshoot another independent being from a knife to the gut.[10]

Indeed, the polyp's lack of a brain and other key organs made it more difficult for people to ascertain if it was an animal. From the perspective of niche academic debate, zoophytes' want of the sensory apparatus of more complex animals has obscured their role in shaping the worlds of humans and dissuaded attempts to entertain their subjectivity or perspective. Though many historians have moved away from seating agency in a conscious, fully rational mind—we are all products of our environment, and all have days when we feel like unreasonable piles of goo, after all—works of animal history still gravitate toward the neocortex-wielding dogs, whales, and wayward pigs of the world. Polyps—unlike, say, a raccoon—seem utterly different from humans. If they have wishes or make decisions, it is through means entirely other than ours. As Spallanzani observed: "Many marine zoophytes are only a kind of jelly. The organization of these animals has not the smallest relation to that of man; plants themselves may be said to resemble him more, because we find sap vessels, utricles, and tracheae in them."[11]

Some eighteenth-century investigators, such as the Scottish naturalist David Skene, however, deemed the capacity to experience sensation as the distinguishing quality of animal life and insisted that the absence of a bounded, brain-like organ did not bar zoophytes from sensing and acting in a surrounding world. "The only proper definition of Animal is an Organized body having sensation," Skene wrote to Ellis in a 1768 letter, identifying the nervous system as "the least ambiguous distinction." Due to this defining trait of sensation, Skene thus identified sertularians, a type of hydrozoan related to corals, as being "as much an Animal as an Elephant," the latter of which some regarded as the nearest animal to humans in intelligence. To the amazement of naturalists, brainless coral polyps could even produce underwater structures that

Figure 1.4. A specimen of brain coral from the late eighteenth or early nineteenth century. (© Sir John Soane's Museum, London, M769; photograph by Lewis Bush)

uncannily resembled a human cerebrum with their "curious windings," a feature that led to these creatures being called brain coral or brain stone (figure 1.4).[12]

Not all natural historians at the time insisted on rigid divisions between animals, plants, and minerals. The French naturalist Georges-Louis Leclerc, Comte de Buffon, for instance, mostly rejected the strict classifications of Linnaeus, specifically describing the freshwater polyp as "the last of animals, and the first of plants." And Linnaeus himself, it should be noted, *did* concede to some mixing of categories when it came to the perennially elusive corals, polyps, and other zoophytes in various editions of his *Systema Naturae*. Schemes like Skene's, which differentiated plants and animals based on the presence or absence of discrete criteria, often crumbled in the face of the empirical multiplicity of nature. For, if sensation set animals apart from plants, what on earth was one to do with a "sensitive plant," such as the Venus flytrap, a botanical curiosity found only in small pockets of the Carolinas? In addition to identifying animals based on traits such as sensation or digestion or spontaneous movement, practitioners like Ellis developed more active litmus tests such as burning the being in question to sniff for a distinctly animal odor. Humans made use of their own animal bodies, senses, and subjective assessments as instruments for ordering these vexing beings.[13]

Despite the difficulties of distinguishing animal from plant at their limits, however, animals did, intuitively, appear unique to many investigators. Even in its sleepiest, subtlest, most basal forms, animality seemed distinctly willful. James Parsons, an English physician and fellow of the Royal Society, explored the line between animal and plant in a 1752 work inspired by Trembley's polyp experiments. Although Parsons found much in common between animals and plants, he insisted a unique "animating Principle" powered humans and other animals, and in doing so he took aim at René Descartes's idea of the soulless animal machine. "How can it be thought," wrote Parsons, "that any Animal is a meer *Automaton,* even tho' it differed from the Vegetable only in Locomotion, which alone is enough to prove a *Volition* in the Creature[?]" Even though the Cartesian notion that animals were nothing more than fleshy clocks was never as hegemonic as often painted, and Descartes's ideas on the matter were at times misrepresented, debates about zoophytes and the precise contours of the animal kingdom further unseated such a view. In the minds of thinkers like Parsons, even the lowliest animals possessed some basic volition—a will that threatened to break down the vision of animal nature as merely mechanical. Parsons was not even troubled by the possibility that this realization might put nonhuman animals on "too equal a Footing" with people, especially when confronted with the rich emotional life of his dog. He remarked, "I had much rather that all other Animals should be intitled to an *Eternity* with me, than that I should be dissolved and annihilated with them."[14]

At a visceral level, then, naturalists tended to view animals as better masters of their destiny than plants, given that they could move, escape predation, and fend for themselves. Their defining trait was self-possession: a sense of self. Oliver Goldsmith described a plant as a "prisoner of nature," unlike an animal, which was capable of "commanding Nature" through animation and defense. Even a lowly polyp could hunt and recoil. Goldsmith believed that all animals had "one power, of which vegetables are totally deficient; I mean either the actual ability, or an aukward [*sic*] attempt at self-preservation." He concluded that "here, I think, we may draw the line between the animal and vegetable kingdoms. Every animal, by some means or other, finds protection from injury; either from its force, or courage, its swiftness, or cunning." Arguing along similar lines several decades later, William Smellie, a Scottish printer-naturalist and influential translator of Buffon, wrote: "Animals will, determine, act, and have a communication with distant objects by their senses. They have the laws of nature, in some measure, at command. They protect themselves from injury by employing force, swiftness, address, and cunning." A polyp moved about and did things for itself. It may have been brainless, but in the eyes of eighteenth-century naturalists that didn't mean it was mindless.[15]

Thus, although polyps lurked below all other animals on the Chain of Being, lacking a brain, bilateral symmetry, and even a face, they were still granted subjectivity. The polyp had the rudiments of cunning, acting as a lively character who shaped its own fate—and the fate of its place in natural histories written by humans. This was especially apparent to Henry Baker, a London naturalist who attempted to replicate Trembley's polyp experiments using animals he received from Martin Folkes, then the president of the Royal Society. Folkes had received some live polyps from Trembley himself and performed Trembley's experiments for fellows of the Royal Society. Baker published his own results for a skeptical British Atlantic audience in the form of a small 1743 volume titled *An Attempt Towards a Natural History of the Polype*—which preempted Trembley's definitive work on the subject, to Folkes's embarrassment. Throughout the work, Baker cast the polyp as a crafty and "tenacious" predator possessing a "most exquisite Sense of Feeling," an animal who "cunningly" outmaneuvered the prey Baker dropped in its vessel. Baker found many of Trembley's experiments with the creature impossible to re-create, such as flipping it inside out; consequently, he regarded the animals themselves as being at times "altogether unmanageable."[16]

Consider how such animals had a hand—or a tentacle—in their own afterlives on the printed page. The polyp's volition and shifting shape tangibly affected its representation in Baker's text, and even Baker's choice of printing technology for immortalizing the animal visually in the book. Baker conceded that the many somersaults and transformations of the polyp dictated his choice to use woodcuts, which are prints made from carved blocks of wood, rather than engravings printed from copper sheets of metal. The sheer number of images required to convince readers of the animal's odd behaviors would have made engravings prohibitively expensive, Baker reasoned, as they would have to be printed from metal plates incised by expert artisans. Moreover, engravings usually stood alone on their own pages since they required a separate high-pressure rolling press, whereas woodcut images could be interspersed with text on a single page, the wooden blocks being the same height as the individual letters of movable type. Baker thought that copperplate engravings would burden the reader with "a good deal of Trouble in turning continually to them: but by being cut in Wood, they [the animals] lye much more conveniently under the Eye in the Places whereto they properly belong . . . though not so beautiful as Copper-Plates." Woodcut printing immobilized the creature for the naturalist's and reader's gaze, integrating illustration and narration of the animal's outlandish nature.[17]

Small woodcuts of the animals richly interrupt Baker's text (figure 1.5). And they do so at a much higher proportion than the average natural history book of the day: nearly half of the pages combine image and text, and some pages contain as many as ten individual images. The images are deceptively simple, however. The polyps appear as thin dots and lines, but woodblock printing is a form of *relief* printing. As opposed to *intaglio* printing methods such as engraving—which involves using a sharp tool to incise a design on a metal plate, filling the resulting grooves with ink, and thus inking onto the page the lines that have been incised—relief printing entails the opposite: cutting away everything that will *not* show as black ink in the final product and inking what remains, much as one would make a rubber stamp. To achieve these simple polyp designs with ample white space, Baker's printer would have had to scoop out almost the entire surface of each block, leaving only a few delicate ridges of wood untouched. The surgical slicing actions Baker performed on his polyps replayed over and over on the wood block, but in reverse. Indeed, I subjected myself and twelve undergraduate students to re-creating some of Baker's woodcuts in the campus printing shop, and we can all attest to the patience, hand cramping, and occasional bloody finger required to spawn a simple polyp from wood.

If living polyps helped mold the labor of natural history in such ways, the many-headed hydra likewise prompted troubling philosophical conundrums, all while posing fundamental questions about the knowability of animals. For example, if naturalists could bring forth new polyps with nothing more than kitchen implements, was this considered a form of creation? And if naturalists were creators, did they still need God? Such questions could lodge a pit in the stomachs of thinkers who sought to find the Creator's fingerprints across the natural world. Some minds, ranging from the lesser-known Parsons to the famed mathematician Gottfried Wilhelm Leibniz, believed animals had souls, a notion tracing back to Aristotle (who divided souls into rational, sensitive, and nutritive). But that presented another chilling polyp puzzle for Baker and countless others. For what happened to the soul when the hydra was split

EXPERIMENT XV.

Quartering a POLYPE.

JUNE, 9, 1743. An *English Polype* being placed conveniently on a Slip of Paper, I cut it entirely through the Body and Head the long way, from Head to Tail: and then, turning the Paper, I gave it another Cut, tranſverſly, acroſs the Body; whereby it became divided into four Quarters, tho' not equal Ones. The Section is expreſſed by the Figure A.

Preſently after the Operation the four Pieces appeared in the Water as (b c d e).

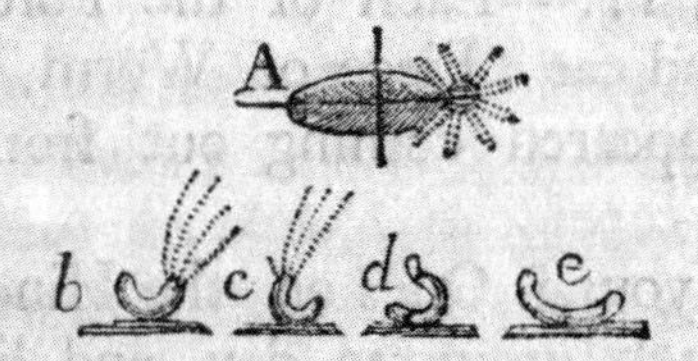

The Experiment was made about four o'Clock in the Afternoon; and, at ten the ſame Night, each of the Fore-Quarter Pieces eat a Bit of Worm, and lay along extended in the Manner of (a b).

The two hinder Parts were contracted, but appeared plump and rounded, as (c d).

June,

Figure 1.5. Woodcuts of Henry Baker's polyp experiments interspersed with text. (Henry Baker, *An Attempt Towards a Natural History of the Polype* [London, 1743]; Biodiversity Heritage Library)

asunder and became two fully functional animals? In Baker's words, "when the *Polype* is divided into several Parts, [and] all soon becomes perfect *Polypes,* where shall we find the Identity of the original *Polype*?" Was the soul indivisible? Or could two souls be made from one? And if the latter proved true, what did that mean for identity, for individuals?[18]

In response, some naturalists decided that certain things couldn't, or perhaps shouldn't, be known. Baker, for his part, declined to answer, wearing his ignorance of the ultimate workings of nature as a badge and balking at those who attempted to mechanically explain every step in every natural process. "'Tis, methinks," he wrote, "a little presuming to restrain the Operations of Nature, or imagine that God has done nothing but according to certain Rules well known to us." Baker, as would be the case with many other British naturalists, believed focusing on the *what,* rather than the *why,* of zoology would make people "probably . . . wiser, and perhaps better." He recommended steering a course of philosophical modesty focused on experiment and knowledge that could be acquired through lived experience and empirical observation. When people pretended "to discover and describe the Machinery whereby, and the manner how these wonderful Effects are performed, which we neither have Senses to discern, nor Abilities to judge of,"—that, in Baker's mind, was the moment when the true threat of ignorance prevailed. For in such a moment, "all is Darkness and Uncertainty, we plunge into an unfathomable Abyss without either Star or Compass to direct our Course, and are in the utmost Danger of Shipwrecking our Understanding."[19]

And thus the very catastrophes that corals caused—shipwrecks—became a metaphor for describing the dangers of attempting to claim total knowledge of animals.

ALL THE WORLD'S A CORAL HUNT

Studying freshwater polyps collected from stagnant ditches was one thing. Comprehending the nature of their ocean relatives, which fanned across the shallows and bottoms and submarine mountains of the world's seas as corals, proved a more daunting task. Observations of freshwater polyps hardly closed the case on ambiguous animals. Rather, they set additional studies of aquatic animal nature in motion. The saltwater kin of the charismatic hydra were stranger, less accessible, and highly variable, situated as they were in deeper waters, bound together in illegible configurations, and capable of turning liquid and light into stone. To look at corals was to look through a glass, darkly.

Their study required distinct constellations of human observers, along with distinct methods. John Ellis, the French physician Jean-André Peyssonnel, and the cadre of other naturalists working with marine zoophytes promoted a seaside empiricism that largely depended on experiences knee-deep in water, especially in imperial colonies.[20]

Peyssonnel investigated corals and worked intimately with coral fishers while at home in Marseille, on the coasts of northern Africa and the Caribbean island of Guadeloupe, and during travels elsewhere in the Atlantic and Mediterranean. Although the sixteenth-century apothecary Ferrante Imperato suspected the corals in his grand cabinet of curiosities might have some animal connection, Peyssonnel has usually received the honor of having confirmed their animal nature. In fact, he recognized coral polyps as such in the 1720s, even before word of Trembley's hydra spread, but his discoveries faced ridicule in French scientific circles for decades until Trembley's work supplied vindication. Ellis, beginning in the 1750s, expanded this line of research in Anglophone scientific circles, illuminating the animal nature of corals and other zoophytes, especially "corallines"—another term of confusion that turned out to encompass some plants and some animals, including calcareous algae. For his efforts, Ellis received the Royal Society's prestigious Copley Medal in 1767. Ellis brought questions about corals and animal nature mainstream, and also away from the mainland. In 1755, on the eve of the Seven Years' War, the Cornish naturalist William Borlase told Ellis in a letter, "You have set all the world a coral hunting, and if we have war or peace, the curious will be at the sea side, I find."[21]

Corals, in essence, asked naturalists to act a bit more like corals: conjoined, coordinated, and cosmopolitan. The volition of corals and other zoophytes, their liquid habitats, and their global distribution all made naturalists' encounters with them dependent on other humans and vast imperial networks. Borlase offered a transnational vision of the collective work of coral study, as these animals spanned not only the warmer waters of the world but also cold northern seas being explored. In fact, more than half of the known coral species on earth inhabit frigid waters beyond the reach of sunlight, forming reefs without the collaboration of photosynthetic partners. Corals were harvested in colonial outposts and various island and littoral environments, stretching from the British Isles, to the far northern Atlantic, to the East and West Indies. Coral science, then, was by necessity a social endeavor. No one person could go it alone. Ellis's reliance on fishers and mariners across the world, for example, allowed him to describe one deep-sea species of sea pen, a type of octocoral, near the North Pole that would not be observed again for more than a century. Corals might, Borlase hoped, offer a chance to transcend politics—though his sunny proclamation papered over the violence and inequality that coral study frequently entailed as well.[22]

Coral polyps generated artifacts that were, quite literally, collectible. Almost anyone could gather fragments of coral that washed ashore, and a good swimmer could pluck corals from shallow-water reefs. Such "marine productions" fit neatly in pockets, though naturalists recommended a wicker basket for safekeeping. Hurricanes also tossed corals, shells, and seaweed "rarely to be met with at other times" onto the shore from less accessible waters, and collecting guides ushered people to the shore after such tempests. Out on open water, mariners reeled in deep-sea coral species stuck to plumb lines or, more proactively, devised various netted submarine contraptions to harvest precious and hard-to-reach red coral (figure 1.6). In a letter to his frequent correspondent Linnaeus in 1768, Ellis salivated at the hunting apparatus of Cook's *Endeavour* voyage, whose tools of coral extraction no doubt disrupted ecosystems. He marveled: "No people ever went to sea better fitted out for the purpose of Natural History, nor more elegantly. They have got a fine library of Natural History . . . all kinds of nets, trawls, drags, and hooks for coral fishing; they have even a curious contrivance of a telescope by which, put into the water, you can see the bottom to a great depth, where it is clear."[23]

Despite the enthusiasm behind this collecting frenzy, the fact that coral polyps were also moving animals rather than mere stationary stones made studying them a vexing process. Soft polyps, when extended underwater, coat the rocky surfaces of the structures they secrete. Corals may have facilitated their study by being readily collectible, but they were still quite difficult to collect and preserve *as animals*. In Trembley's experiments, the jelly belly of the polyp was the unambiguous object of study. But stony corals were both artificer and artifact: a calcium carbonate skeleton fashioned by saltwater polyps could be stored in a cabinet as a curious thing without any obvious trace that delicate animals had produced it. Collections skewed toward what could be dried and shelved—and, as a result, toward forms that resembled stones and, superficially, plant-like ramifications. Unlike their contacts with the hydra, naturalists needed to reckon with the calcified substances and elaborate objects that coral polyps produced. And then, despite appearances, they had to get back to the soft-bodied animal behind it all.

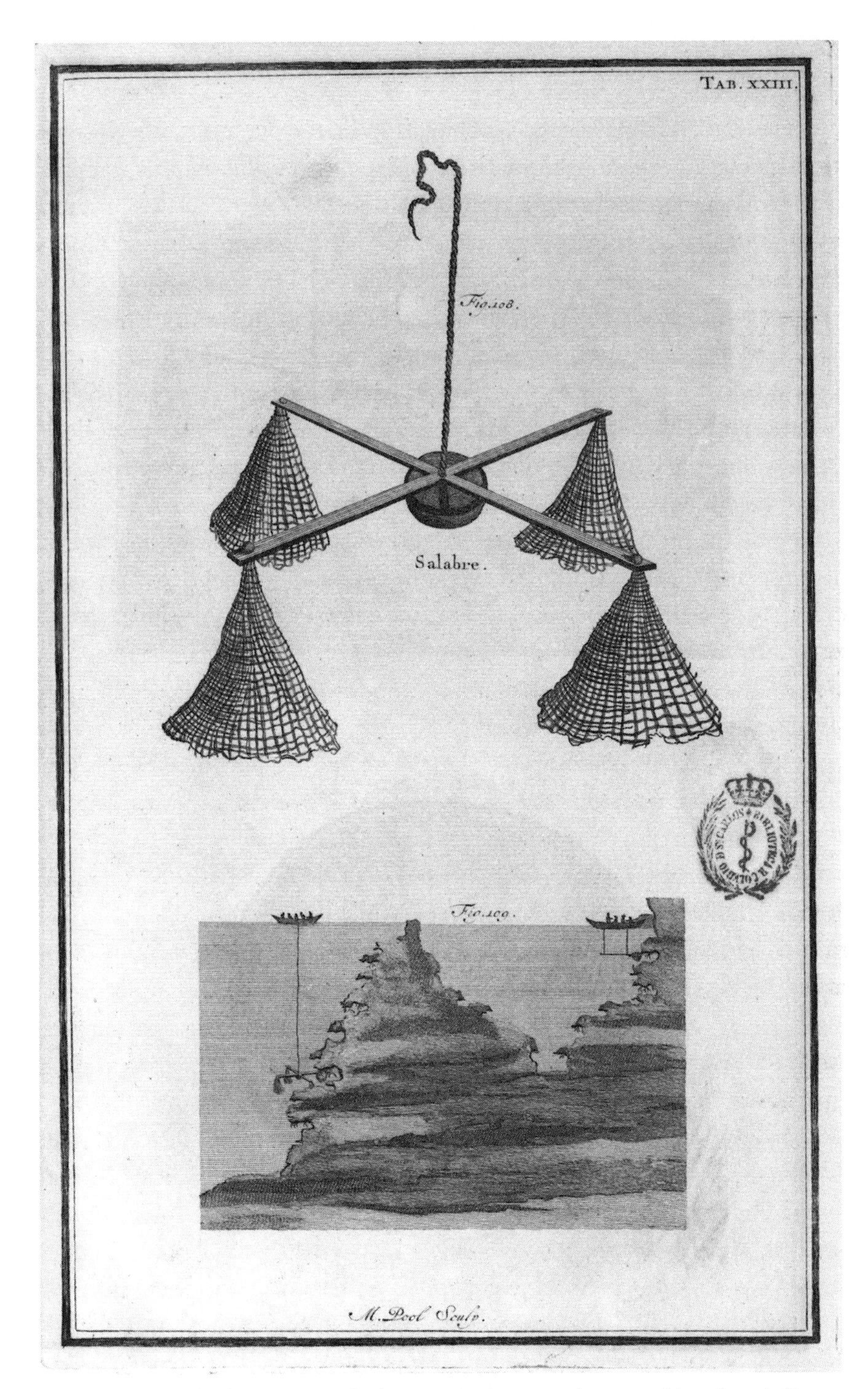

Figure 1.6. Two netted contraptions for harvesting precious and inaccessible red coral. (From Louis Ferdinand Comte de Marsilli, *Histoire Physique de la Mer* [Amsterdam, 1725]; HathiTrust Digital Library)

For example, at the approach of motion, coral polyps quickly retracted into their cup-like skeletal housings, called corallites, making these animal communities challenging to preserve in a manner that made them visible as anything other than inert substances. The volition that partly defined corals as animals also made them hard to prove as such. George Humphrey's handwritten book of collecting instructions suggested carrying corals home in buckets of sea water and watching them closely until, "after some time you will see the little Creatures put forth their Arms or feelers in search of prey; at which instant you must immediately take them out & plunge them into spirits; this must be done with a quick motion, otherwise the Animals will retire into their Cells and consequently will be hid." The movements of these animals taxed the naturalist, forcing him to keep vigil with the polyps and then work dexterously to preserve their animal features, violently suspending them in a grave of liquor. Dicquemare described allied anemones as "almost impossible to keep even after death." As opposed to the soft tissue of animal bodies, it was comparatively simple to preserve the intricate limestone skeletons slowly accreted by polyp communities.[24]

Given the challenges of preserving ephemeral polyps, coral investigators conducted their work at the seashore or, in some cases, below the waterline. They thought in, through, and against water as a medium, huddling in boats with coral fishers and taking in the briny air in Peyssonnel's case, or in Ellis's, bringing skilled artists and so-called aquatic microscopes, like those designed by the optician John Cuff, straight to the water's edge. In a letter to Ellis in 1764, the physician William Brownrigg observed that "the most tender, beautiful & curious" zoophytes grew in the ocean "beyond the low water mark" and in hollows "where they cannot be easily dredged up." To obtain "perfect specimens," he recommended a diving bell—a submersible chamber, usually manned by hired hands or enslaved divers rather than elite naturalists who would not deign to get wet. Water was never a transparent window, for it had its own properties, affordances, and limitations that shaped the study of marine life. William Watson, who presented the Royal Society with translated digests of Peyssonnel's more-than-400-page argument that corals were animals, deemed "the difficulties, which there are, in getting from the bottom of the sea its productions" as the reason why the study of corals had "been hitherto very imperfect." The account also concluded that the small, buffeted boats used by coral fishers destroyed "that tranquillity of mind" needed to contemplate animal nature, and thus only someone well-traveled and well-versed in the "fatigues of the sea" could truly understand the ocean's inhabitants.[25]

Yet even those committed to working with water still might fail to observe zoophytes in motion. In a series of increasingly frustrated letters to Ellis, the cartographer Murdoch Mackenzie relayed his continued difficulties, after several "hot Months . . . by the Sea Side," in viewing the animal activity Ellis had witnessed in corallines. Immediately after receiving specimens from a local fisherman, Mackenzie submerged them in white earthenware filled with seawater for an hour. He watched and waited, observed the specimens with several different microscopes, and followed Ellis's epistolary coaching. But it was no good. Mackenzie's forbearance earned no reward, and he could not "observe the least appearance of Life or Animal Motion." Many of these creatures failed to cooperate and reveal their nature in the manner naturalists expected, and the delicate water conditions required for their survival made them all the more elusive. Extended time at sea did not necessarily produce the intimate conditions for understanding corals, either. Joseph Banks wrote of corals in the Great Barrier Reef, especially *Tubipora musica,* the organ pipe coral (which fabulously resembles its name): "I have often lamented that we had not time to make proper observations upon this curious tribe of animals but we were so intirely taken up with the more conspicuous links of the chain of creation as fish, Plants, Birds &c &c. that it was impossible." Even when corals did reveal themselves in simulated aquatic conditions, to diagnose them as animals remained far from straightforward. Count Luigi Ferdinando Marsigli, an Italian nobleman, placed a live specimen of red coral in a glass bowl of saltwater, as depicted in his *Histoire Physique de la Mer* in 1725 (figure 1.7). But when the white polyps finally, quietly appeared, he couldn't unsee them as brilliant, unfurling blossoms: utterly botanical.[26]

What was more, the watery abode of coral erected a potentially deadly barrier between scientist and specimen. Churning currents, predatory sharks, painful pressure at depth, and even poisonous zoophytes all threatened human life. That alone could serve as occasion for privileged naturalists and philosophers to send mariners, fishers, and Indigenous or enslaved divers to assume such risk instead—although the historian Kevin Dawson has shown that the European conception of a lethal ocean was foreign to many non-Western communities in the early modern period, particularly among Africans, who "wove terrestrial and aquatic experiences into amphibious lives."[27]

Figure 1.7. Polyps of red coral extended in a vase filled with seawater. The petal-like tentacles and apparent "blossoming" of the specimen suggested to Count Luigi Ferdinando Marsigli (sometimes Marsili or Marsilli) the botanical, rather than animal, nature of coral. (From Louis Ferdinand Comte de Marsilli, *Histoire Physique de la Mer* [Amsterdam, 1725]; HathiTrust Digital Library)

Nevertheless, the demands of coral and water brought bitter stratifications in colonial society to the surface. Alexander Garden, a Scottish physician based in Charleston, South Carolina, wished to provide Ellis with zoophyte specimens. But in March 1755 he wrote with vitriol of the fishermen he relied on to perform the task in his stead: "Most or indeed all of them are negroes, whom I find it impossible to make understand me rightly what I want; add to this their gross ignorance and obstinacy to the greatest degree; so that though I have hired several of them, I could not procure any thing."[28]

From Garden's letter alone, the enslaved status of these men remains uncertain given the presence of a modest community of free people of color in Charleston at the time. In either case, they may have obfuscated with a purpose; Garden's demeaning tone doesn't suggest he was pleasant to work with, and, as we will see in the case of his fish studies, this wouldn't be the only time Garden would blame his shortcomings in natural history on people of color. But Garden's commentary does crucially underscore how collecting coral and arranging the natural world ran parallel with attempts to solidify racial hierarchies among humankind. In the process, coral's potential to transform intensified: objects harvested by marginalized laborers could be assigned racialized meanings, as when, in 1776, the collector George Humphrey described the oft-admired brain coral as "Negroe's head." Through such a charged comparison, Humphrey subsumed people of African descent into nonhuman matter, though natural history itself would have languished without the expertise of people it frequently categorized, vilified, and enslaved.[29]

For all of water's challenges, as a medium it could still prove a vital tool for comprehending the in-betweens of the natural world. The star turn of corals and polyps sent ripple effects across studies of marine life more generally, as in the work of Reverend Griffith Hughes, rector of Saint Lucy's parish in Barbados, who wrote *The Natural History of Barbados*, published in 1750. Although other sections of the text reveal that Hughes sent enslaved collectors to do his natural historical bidding, he ostensibly used the properties of water for himself to suss out the nature of "animal flowers" in a seaside cave that remains a tourist site to this day. Tube worms and yellow anemones line the floor of the cavern like daffodils. Yet whenever Hughes tried to "pluck" one of the creatures, as he would a flower in June, it contracted before he could make contact. Caught mentally "for some time in Suspense" like a specimen in water, Hughes had to ask: did the organism, which mightily resembled a plant, simply buckle under the force of the water, which he knew to be "eight hundred times heavier than Air"? Or did its motions imply volition, sensation, and animal life? After another visit to the site, Hughes concluded these were animals, and not plants, when their "seeming Leaves" closed in on prey like forceps. Word of the astounding creatures soon spread across Barbados. The cavern became a must-see destination, but visiting entailed trekking unwanted through the property of a local man who, in a fit of rage, clambered to the cave and attempted to kill all of the creatures with an iron rod. They rebounded from the attack, suggesting to Hughes that "a latent Principle of Life can be preserved, after the whole organic body is torn to pieces."[30]

Although Hughes initially thought his creaturely discoveries too "chimerical to mention," the by then well known propensities of polyps emboldened him to put such observations to paper. Corals and polyps, obscured and distorted as they were through the scrim of the marine, inspired additional inquiries into the intractability of the natural world. When conveying his observations to the Royal Society, Hughes ceded ground to the illegibility of these confounding beings, concluding "we cannot, without the highest Arrogance, presume to prescribe Limits to the Power of the ALMIGHTY, who, for wise Ends, sometimes hides his Works in such Darkness, as to be concealed from the most exalted human Knowledge."[31]

POLYP POLITICS

In a letter from 1769, Linnaeus put Ellis's coral research on a par with the European conquest of the Americas—an apparent compliment, delivered in natural history's love language of flattery. "Whenever I look at those beautiful figures of Zoophytes which you have sent me," Linnaeus wrote Ellis, "an anxious wish arises that I may live to see their descriptions from your pen. Your discoveries may be said to vie with those of Columbus. He found out America, or a new India in the west: you have laid open hitherto unknown Indies in the depths of the ocean." Linnaeus's comparison made explicit the colonial dimensions of coral science. Zoophytes brought people into volatile cultural collisions the world over. We saw this quite clearly in the reliance of Alexander Garden, Griffith Hughes, and other naturalists on collectors of color, as well as in James Cook's near demise by the hands of the Great Barrier Reef, which delivered him into a charged confrontation with the Guugu Yimithirr people on northeastern Australia's coast. Indeed, corals could be more comprehensively studied in the eighteenth century due to aggressive imperial expansion that made possible their global collection and comparison. Yet it was also the very makeup, structure, and shifting being of corals that allowed their study to lapse into social and political commentary. Inquiries into these animal aggregations became a de facto investigation of communities, and thus of colonialism itself.[32]

The one-but-many nature of corals made these animals both alluring and confounding to naturalists. Were they one superorganism? A collective of individuals? Both? Neither? Animal groupings like corals, in fact, first came to be described as *colonial* in the Enlightenment era. Scientists today describe assemblages of coral polyps, bacteria, birds, and other close-knit organismal groupings as "colonies" without batting an eye. It approaches a formal term of art, a basic semantic fact of biology. But the metaphor is just that: metaphorical, both far from obvious and, in this case, the product of a specific historical moment. In the eighteenth century, coral configurations asked those carrying out the twinned work of coral study and colonialism—whether in Europe and the Mediterranean, the Americas, the cold northern Atlantic, or, toward the end of the century, Australia and the Pacific—to rethink the nature of empire, community, and belonging as they observed such beings. So-called colonial organisms and their complex group dynamics, in turn, offered natural case studies for conceptualizing and administering imperial populations and colonies. Their ambiguity made corals particularly available for colonial dreaming.

Much early modern usage of the word "colony" derived from the Roman *colōnia,* referring to a settlement of Roman citizens, especially if they were occupying new territory for the empire. Samuel Johnson's *Dictionary of the English Language* in 1755 cited that Latin definition in deeming a colony either a "body of people drawn from the mother-country to inhabit some distant place" or "the country planted; a plantation." The word's application to nonhumans witnessed an uptick in the eighteenth century as Britain expanded its own empire. In 1792, for instance, the surgeon John Hunter used language reminiscent of the Roman concept to write of bee "colonies" as extensions of the mother country: "they breed to form a colony, which is to go off from the old stock, in order to set out anew." Oliver Goldsmith painted polyps as spectacularly colonial given their penchant for infinite regress, writing: "every polypus has a new colony sprouting from its body; and these new ones, even while attached to the parent animal, become parents themselves, having a smaller colony also budding from them." As the century progressed, "colony" took on specific zoological usage to denote an "aggregate of individual organisms . . . that are physically linked together to form a connected structure," as in the case of the polyps of a coral community, knit together by soft coenosarc tissue. In notes appended to his poem *The Botanic Garden,* originally printed in 1791, Erasmus Darwin even lent zoophyte "colonies" a historical dimension, describing the formation of deep reef structures as follows: "A colony of . . . madrepores [stony corals] . . . lived and perished in one period of time; in another a new colony of either similar or different shells lived and died over the former ones . . . and thus from unknown depths to what are now the summits of mountains the lime-stone is disposed in strata."[33]

In bringing this colonial lens to corals and other zoophytes, naturalists began to study their social dynamics, especially their coordinated movements. Some corals are solitary, like members of the genus *Cycloseris,* which live as singular, disc-shaped polyps the color of psychedelic trips. Many other species go it alone at some point in their life course. Most corals, however, suture themselves together in entangled and ontologically muddy groups. Peyssonnel's translated tract, for example, noted that the polyps of many scleractinians "seem to constitute one and the same body." Ellis marveled that one "many-bodied animal" under his study contained "so many different bodies united in one; acting like so many sets of hands, placed in [the] form of a circle, collecting food, each for a mouth in the centre, to convey nourishment to so many stomachs."[34]

Unlike a beehive, with a queen at its helm and a hierarchy of workers and soldiers, individual coral polyps seemed radically equal and leaderless, yet still synchronized and forming one coherent whole. John Albert Schlosser, a correspondent of Ellis's, noted of a specimen: "their mouths, and their motions, were perfectly the same, in every one individual." Many coral groupings are indeed made up of genetically identical, or nearly identical, polyp clones, numbering frequently in the thousands and occasionally the millions. In form, these bodies conjured Thomas Hobbes's many-peopled leviathan, only without a ruling head at the top (figure 1.8). William Whewell would later make this resemblance explicit, writing in the nineteenth century that Hobbes's body politic was "a kind of polyp, as appears by his frontispiece." The structure of corals cut to the heart of political philosophy and the social contract: how was one to reconcile the one with the many, the *pluribus* with the *unum*?[35]

The clustered, bustling nature of coral communities conjured various and often contradictory visions of politics. Many times, they held up a societal mirror to the observer. As Alexander Garden contemplated a 40-pound specimen that was likely brain coral, he read it as a kind of map and told Ellis that the sample was "all divided into kingdoms and commonwealths by high ridges running in various directions above the common surface about half an inch." Ellis's own notion of the zoophyte "colony" was pliable. In some cases, he used the word simply to show the aggregate and coordinated nature of these animals, much like biologists might do today, presuming it to be a neutral term. For instance, when noting that sponges exhibited that "different Species of Resistance" from inanimate matter, Ellis referred to the animal structure as "the whole Colony" without further elaboration. At other times, however, he applied a more politicized valence: writing of corals, he called their animal operations "the Vigour of the Republic," pitted against encroachments by other sea creatures and animal nations.[36]

Through his more charged descriptions, Ellis explicitly revealed the role

Figure 1.8. Thomas Hobbes's sovereign and subjects, making literal the notion of a body politic. (Frontispiece [detail], Thomas Hobbes, *Leviathan* [London, 1651]; Wikimedia Commons)

of colonization, human or otherwise, in violent expansion and displacement. Writing with an air of inevitability, he claimed that the waters of warmer equatorial regions abounded "so much with Animal Life, that no inanimate Body can long remain unoccupied by some Species." Indeed, he insisted that all surfaces, from underwater plants to shellfish, were vulnerable to becoming "the Basis of new Colonies of Animals, from whose Attacks they can no longer defend themselves." Marine animal nature, in this view, had an inherent taste for conquest—especially in latitudes that were likewise hotbeds of human subjugation.[37]

Even Ellis's essay that won the Copley Medal in 1767 connected the study of zoophytes with colonialism in its very title, "An Account of the Actinia Sociata, or Clustered Animal-Flower, Lately Found on the Sea-Coasts of the New-Ceded Islands." It contended with the natural history of Caribbean islands recently ceded from France to Britain through the Treaty of Paris in 1763, which ended the Seven Years' War—islands including Dominica, for which Ellis would eventually serve as an official colonial agent. In the prizeworthy piece, he attempted to tease out the existential conundrums inherent in colonial arrangements, as he noted in the first-ever description of what is now commonly called the green sea mat: "an animal compounded of many animals has not a very philosophical sound. But it is well known to those, who understand the nature of zoophytes; that there are many kinds of these animals." The outcome of the Seven Years' War, often regarded as the first global war, granted the British Empire immense swaths of land in the New World and forced imperial administrators to manage and incorporate a radically diverse population spanning French residents in Canada, French and enslaved residents of the newly ceded Caribbean islands, Spanish communities in Florida, and Native nations and empires throughout all of these territories, including the Ohio River valley. The nature of aggregate animals offered potent political metaphors at a time of major imperial reconfigurations across multiple continents. Creatures that proved quite dangerous to colonizers in practice also aided in conceptualizing colonialism.[38]

On its surface, the self-evident homogeneity of a coral community seemed at odds with the hierarchies naturalists and imperial administrators alike sought to impose on humans and nonhumans. Yet on the ground in North America, the sameness of polyps suggested a startling political blueprint to the polymath Benjamin Franklin. In his *Observations Concerning the Increase of Mankind,* originally penned in 1751, Franklin compared the British Empire to a polyp that could sever and send forth its limbs to watch its colonial progeny regenerate anew. "Thus," wrote Franklin, "if you have Room and Subsistence enough, as you may by dividing, make ten Polypes out of one, you may of one make ten Nations, equally populous and powerful." Written in protest of restrictions on colonial manufacturing, the tract painted a picture of the utility and fecundity of the American colonies to the British Empire. Franklin noted the colonies' capacity for swift population growth, since the American continent had land in "plenty" (here he discounted the nuances of both Native sovereignty and dispossession). With the roving, regenerating polyp as inspiration, Franklin offered a racially homogenous vision of what he thought such an expanding populace should look like. In particular, he feared that "the Number of purely white People in the World is proportionably very small"—a trend he hoped to reverse. "All *Africa* is black or tawny," Franklin wrote, cataloguing humanity in a natural historical register, based on sweeping generalizations of morphology. "*Asia* chiefly tawny. *America* (exclusive of the new Comers) wholly so." Why, asked Franklin wryly, in the process of "*Scouring* our Planet, by clearing *America* of Woods, and so making this Side of our Globe reflect a brighter Light to the Eyes of Inhabitants in *Mars* or *Venus,* why should we in the Sight of Superior Beings, darken its People?" Franklin outlined his disturbing dream of a uniform white empire spreading round the clearcut earth like a budding chain of milky polyps, using nature itself to naturalize the notion that "such Kind of Partiality" to one's own complexion was manifestly "natural to Mankind."[39]

Although polyps can hardly be blamed for Franklin's racialized rhetoric, these concepts and metaphors did not emerge from thin air. As captured provocatively by the historian Bathsheba Demuth, "The state of nature conditions the nature of the state." The supple, blank, and bunched nature of corals and polyps influenced imperial fantasies, creating the conditions of possibility through which naturalists could twist the natural world toward their own ends.[40]

And yet, out of the land of metaphor and into the land of the material, coral polyps also had the power to unwittingly level human hierarchy, if only fleetingly. When Cook's *Endeavour* crashed into the Great Barrier Reef, the ship's passengers, regardless of status, were forced to pump water in a frantic fight for their survival. As described by Sydney Parkinson, "every man on board assisted, the Captain, Mr. Banks, and all the officers, not excepted." Banks expressed some shock that the lower ranks did not resort to mutiny at this moment, "contrary to what I have universaly heard to be the behavior of sea men who have commonly as soon as a ship is in a desperate situation began to plunder and refuse all command." But desperate times apparently called for desperate measures from both ends of the hierarchy. Tiny animals, working in concert to build a reef, turned a floating pack of saltwater humans into pumping animals momentarily void of stratification, and without much separation from the other animals, either. As the century marched on, naturalists had to face the reality that corals proved as adept as people at colonizing the globe.[41]

A NEW WORLD ORDER

In a deeply hierarchical world, mired in deeply hierarchical worldviews, corals turned cherished human assumptions about the natural order on their head. If freshwater hydras raised troubling philosophical questions about the boundaries of life and animality, never mind the nature of the soul, saltwater polyps sparked unsettling debates of a different sort to armchair naturalists. The notion that the smallest, softest, lowest-ranked animals on the Chain of Being fashioned complex and enduring underwater structures in the form of reefs and other coral creations made certain thinkers distinctly uncomfortable. For some, it exposed fissures in their view of God's hand in nature. One revealing debate on this paradox took place through the channels of the Royal Society. In response to the manuscript that Peyssonnel sent suggesting the animal nature of corals, a physician, Anglican clergyman, and noted antiquarian named William Stukeley submitted a lengthy takedown that was read at the Society in 1752. To Stukeley's consternation, the society never printed it in the *Philosophical Transactions,* instead opting to publish an abridged version of Peyssonnel's tract. Historians have largely ignored Stukeley's argument, both because it is in manuscript form and since he ostensibly "lost" the debate. But we cannot dismiss the dismissal: to do so misses key theological and philosophical arguments of the time, as Stukeley's statements echo claims made by a range of naturalists, physicians, and philosophers. This seeming dead-end of scientific thought crystallizes *why* corals and liminal animal volition rankled many a human investigator.

Stukeley specialized in the emerging field of archaeology, and notably the ruins of Stonehenge, believing it to be of druid origin. Like many fellows of the Royal Society, however, he was an intellectual with interests as diverse as gardening, physics, freemasonry, and elephant anatomy. Born in Lincolnshire in 1687, he counted luminaries like Martin Folkes, Hans Sloane, and Isaac Newton among his friends; in fact, the earliest recorded instance of Newton confiding his falling apple anecdote was to Stukeley as the two sipped tea in Newton's garden. And, as would be the case for so many other Enlightenment-era naturalists, Stukeley was electrified by the question of corals.[42]

Stukeley conceded that saltwater polyps themselves might be of an animal nature, or at least some mixture of animal and plant. But he found it absurd to suggest that polyps, akin to "those we observe in our own fresh waters," produced marine coral formations, given the agency it would grant "so helpless a creature" and potentially wrest from God, in return, in what he envisioned as a zero-sum game. Stukeley saw no reason to think polyps *made* their rocky habitations, just as it would be an error to assume that "worms that borough in our ship planks can be called the shipwrights that built the ships." Polyps were parasites of already extant structures, in Stukeley's mind, not providentially gifted creators. "Instead of bestowing this alm[ighty] power on the very lowest, least & most impfect [imperfect] species of creatures," wrote Stukeley, "I see much reason to admire the wisdom & goodness of the creator, who has thus disseminated his beautiful groves of corals & its kindred along the bottom of the ocean & its seas, like bushes on a wild common." In Stukeley's opinion, God created the corals, which were "real plants," and the polyps simply followed in step by making those God-given corals their homes.[43]

Stukeley's vision of submarine gardens proved compelling amid a historical zeitgeist that conceived of the globe as a coherent and largely consistent unit, efficiently and abundantly furnished by God with all of its needs. The universe exhibited plenitude, continuity, and gradation, as articulated by Arthur Lovejoy in his landmark work on the intellectual history of the Chain of Being—indeed, so much so that many philosophers surmised that other planets must be populated like the earth, in a so-called plurality of worlds. The eighteenth century also saw the expansion of imperial botanical gardens that showcased the flora of the globe in microcosm. Gardening was big business, planting an act of power. Stukeley, in addressing his tract to the Quaker gardener-merchant Peter Collinson, insisted that God planted corals to "adorn the plains of that great liquid element, incompassing the globe: thus to render every parcel of the creation habitable: to add to it the immensity & infinity of the objects of his paternal care." He viewed all coral polyps as instinctive and more or less homogenous, and thus he could not entertain the "mean" idea that one brainless animal made the seemingly infinite variety of "very elegant forms" of corals, a "variety as wonderful as their beauty," which he witnessed on display at various collections of corals across London.[44]

For Stukeley, the question of coral's origins was in part a problem of geometry. He insisted that the supposed power of polyps to create such variable forms—which differed from one another in "a thousand ways," with compositions as diverse as brain coral, sea fans, and staghorn coral—found no analogy in the honeycomb of bees, which made cells of repeating shape and size, or in spider webs, with their "invariable geometry." Repeating building blocks in nature were tenable as animal productions since they implied blind, repetitive, "divine instinct." But Stukeley deemed complex fractals and the seemingly limitless variety of coral bodies too complex to have been produced by animals—especially by animals who were themselves less structurally intricate than bees or spiders. James Parsons, a sometimes defender of the miraculous powers of polyps and other creatures, supported his friend and colleague Stukeley in this case, calling polyps "little, poor, helpless, jelly-like animals."[45]

Stukeley furthered his arguments through close inspection of several dried specimens of corals and corallines. He admitted, "I have never been at the bottom of the ocean, at coral-fishing" but maintained it mattered not, as specimens, he believed, could speak for themselves. He assured his audience: "autopsy suffices in the case." It is perhaps no accident that Stukeley, as a student of ancient human monuments, would place unwavering faith in material artifacts, be they cracked and dried ruins of human civilizations or corals. Unveiling a dried specimen from Gibraltar to the Royal Society, Stukeley contended that the object itself would answer the polyp question through its own, self-evident, "cogent argument." Stukeley urged those present to look closely at it, "with the naked eye, or rather with the microscope," to observe the "little globules" spread across its surface. Most contained a small depression or cavity, which Stukeley argued were the individual resting places of polyps (that is, corallites), ingeniously designed by God for "millions of inhabitants." But he also revealed that not every globule had a perfect indentation for a polyp, proving, to him, that these cups were animal *destinations,* where polyps had seized upon a globule and caused an indentation—not animal *designs.*[46]

In basing his arguments on dried specimens of coral that had been removed from their ocean origins, however, Stukeley was indeed facing the age-old dilemma that coral appearances can deceive. He even wrote in his diary: "To deny them being vegetables is to put a cheat upon our eyes." And yet, corals have always cheated human perception. When independently working to observe the same specimen under a microscope with Parsons, Stukeley noted that a "stem" broke off. They considered it a boon, for "the fibrous & woody texture of it appeared so notorious, that it is impossible to deny it, to be intirely a plant incrusted over by a stony case." In other words, the internal structure revealed by the accident seemed to support their theory of a botanical (and thus, God-planted) origin. The frequent breakage of dry specimens, however, made these skeletons unreliable narrators of their history, since Stukeley's own arguments rested on close inspection of slight irregularities in a specimen's finest features.[47]

What did it matter to Stukeley, or Parsons, or other naturalists, if tiny animals generated vast coral gardens in the sea? Imagine suddenly learning that butterflies, not people, planted all the trees in Central Park; or that every flower, every blade of grass, every bush in the hedge outside the window of your childhood room, which you swore you saw your mother dutifully tend every summer, along with each rock in the yard, even the one you selected for your collection that was impossibly shaped like a human hand, had been excreted by worms. Such surprise would be felt as acute disruption in Stukeley's world, where a gnawing gap between the minuscule and the monumental—between the "soft gelatinous weakly" polyp, to borrow another slight from Parsons, and the animal's creatively rendered stone habitat—threatened God's own dominion and artistic supremacy. Stukeley found animal power subversive, even borderline sacrilegious. He classed himself as "very unwilling to take this task out of the hands of the sovereign planter, & give it to the helpless polypus." Because of the polyp's low station on the Chain of Being, the notion that it might produce "the immense variety & elegance of those coral bodys" and act as a marine landscaper seemed poised to disrupt the natural order. As corals took in sunlight, grasped tiny zooplankton with their tentacles, and secreted layers of limestone, they raised worlds—worlds that shook the bedrock of human certainty.[48]

Although plants occupied a rung below animals on the Chain of Being, some naturalists found it more appropriate for an advanced plant to produce intricate forms than for the lowermost animal to do so. In a similar vein, others feared that zoophytes revealed flaws and mismatches in the system. Murdoch Mackenzie told Ellis in 1756, "I can't imagine why a Small Marine Plant toss'd about with every Wave, shou'd be endued with Animal life as you have proved to Conviction, while a Noble Oak which . . . appears one of the most beautiful, most Majestic, and most useful Species of the Creation, shou'd be allowed a Vegetative life only." As his comments reveal, people valued animation and animal life in themselves as higher states of perfection but sometimes cringed at their allocation across the natural world. For other Anglophone naturalists, however, the productive capacities of the smallest animals could be a cause for celebrating the limits of human industry and understanding. In the 1780s, the American naturalist William Bartram admitted that although animals such as coral polyps could never fabricate "a compleat Ship, a Watch or Clock, Mariners compass, Iron or Steel," it was likewise "equally impossible" for humans to make honeycombs, spider webs, or coral reefs. Bartram concluded that a human being "no more than any other Animal possesses a creative power" and "can at most only work upon or modify Matter already created to his hands & so can most other Animal[s] in some degree or other."[49]

The stakes of coral power took on increasing urgency as the eighteenth century progressed. Research by Johann Reinhold Forster, which would be built and expanded upon, coral-like, by Matthew Flinders and Charles Darwin in the nineteenth century, suggested polyps made more than the endless underwater tableau of coral shapes: they also raised entire reef structures that emerged from ocean depths. Corals could even generate islands. In creating submarine monuments, animals changed the literal course of empire, demarcating the paths maritime expeditions could follow and forcing humans to recognize that the lowest among them had consequential power. Blessed were the meek, for they created the earth.[50]

Given that so-called coral rock wrecked the hulls of ships and likewise provided a poor surface for anchoring, ship crews changed their plans or even abandoned visiting certain shores once they noted the presence of coral. Others were not so lucky. Reefs claimed vessel after vessel in the broader early modern period, forming ship graveyards that became further encrusted by coral—negative underwater museums made by animals, but for purposes less legible. Oliver Goldsmith described coral as "the work of an infinite number of reptiles of the polypus kind, whose united labours were thus capable of filling whole tracts of the ocean with those embarrassing tokens of their industry." In this context, eighteenth-century use of the word "embarrass" specifically meant to hamper or impede. Observers also realized that coral would soon overtake man-made objects that sank to the ocean floor, as would happen to relics of the *Endeavour* voyage that shipmates heaved overboard to lighten the vessel as it gasped on the Great Barrier Reef. The English physician John Woodward observed: "The Coral found affix'd and growing upon wreck'd Ships, lost Anchors, and various other artificial Bodies, that are daily dragg'd up out of the Sea, affords a Demonstration that Coral continues to be form'd to this Day." The action of corals on the human world, and the interface of the natural and artificial, then, offered a means of thinking about whether these animals changed over time and formed a history of their own, rather than being created in one fell swoop at God's command.[51]

"Great effects seem to be sometimes wrought by comparatively small means." Those were the words of the New York naturalist and U.S. senator Samuel Latham Mitchill decades later in 1804, as he observed the same gap between tiny polyp and towering coral witnessed by Stukeley, which caused the latter such metaphysical grief. Rather than horror or unease, the outsized output of coral polyps inspired awe for Mitchill. He also helped his wife, Catherine, amass a cabinet of New World corals, which they called her "Bahama Collection." Samuel told Catherine in anticipation of his return home: "When you get all these Productions of the Ocean, you will look like Amphitrite the Goddess of the Sea, and I believe I shall assume the character of Neptune, and come along as soon as ever I can get loose from my moorings here and make love to you."[52]

In their love letters, the Mitchills continued to consider questions of animal power that had been debated in the previous century, but to different effect. Samuel carefully attended to the "work" and "incessant Industry" of coral polyps. Marvel at the labor of polyps—often explicitly labeled "agency"—would remain a theme throughout the nineteenth century. For instance, in 1809, George Shaw similarly wrote that "the Coral tribe, however insignificant it may at first appear, is one of those powerful engines in the hand of the Author of nature which can produce the most stupendous effects from the most seemingly weak and unpromising agents." Samuel went on to explain how the microscope had revealed that tiny animals, which "live and procreate without the knowledge of Man and beyond the limits of his Empire," made calcareous mountains in the sea, many of which "here and there emerge from the water as Bermuda and Barbados seem to have done, and form Islands of Limestone." As Samuel's comment shows, corals acted beyond the reach of human dominion and knowledge to generate, in some cases, the physical stepping stones of empire. In Stukeley's eyes, giving lowly animals the creative power of terrain formation would have diminished divine agency. For the Mitchills, the productivity of coral polyps further cinched the mightiness of the Creator's works.[53]

The meanings of coral would continue to multiply after the eighteenth century. Whereas earlier characterizations of coral "colonies" were apt to stress violent conquest and territorial expansion, coral communities would eventually lead people to promote the leveling of social hierarchy. Scholars such as Danielle Coriale and Michele Navakas have studied the subsequent nineteenth-century use of coral—by poets and political philosophers, including Karl Marx himself—to advocate for collectivity, women's domestic labor, and socialism. As for political theorists of late, James C. Scott linked polyps and peasants in an eloquent comparison: "Everyday forms of resistance make no headlines. Just as millions of anthozoan polyps create, willy-nilly, a coral reef, so do thousands upon thousands of individual acts of subordination and evasion create a political or economic barrier reef of their own." Over time, the indeterminate biology and social structure of corals have allowed them to be mobilized toward very different ends—though coral, much as man, has always been a political animal.[54]

Corals indelibly shaped, and also irrevocably foiled, human projects to understand and categorize animals. Conclusive knowledge of such beings remained elusive. By the end of the eighteenth century, many practitioners decided that the precise line separating animal from plant might be beyond the grasp of human knowledge, undermining the notion of a legible hierarchy of being. In 1785, William Frederic Martyn threw traditional systematics out the window when he decided to present creatures in his animal "dictionary" alphabetically instead of taxonomically, given the insurmountable haziness of nature. He defended his decision to "emancipate" his text from "system" with an appeal to the "sublime disorder of Nature herself, too prolific to enumerate or arrange her productions; and the essential variations between the most celebrated Naturalists, who confound while they attempt to explain." William Smellie believed the distinction between plant and animal surpassed "the limits of human capacity," despite how simple it might seem to tell a dog from a dogwood. Growing, living, and feeling were "only the passive properties of animals," in Smellie's view, failing to capture their animate essence: those "instinctive, intellectual and active powers which exalt the animal above the vegetable." He concluded that "definitions, when applied to natural objects, must always be vague and elusory. We know not the principle of animal life. We are equally ignorant of the essential cause of vegetable existence. It is in vain, therefore, to dream of being able to define what we never can know."[55]

Most twenty-first-century biologists identify corals as animals without giving it a second thought. But at the level of nomenclature, corals remain classed as anthozoans—as "flower animals." The zooxanthellae that lurk in their tissues are similarly fuzzy. Many a scientist would now call them protists, a catch-all category for any eukaryote that is not an animal, plant, or fungus. It could well be our own contemporary version of "zoophytes." Even the term "zooxanthellae" has come to be viewed as outmoded by some practitioners given the word's capaciousness and competing definitions, not to mention the multiplicity of other bacterial, viral, and fungal symbionts living in and through coral. Nature's sublime disorder reigns still.

Corals forced early naturalists to confront the nature of animals and their volition, the nature of communities and individuals, and the nature and consequences of animal power. Animals were not merely a springboard for human concerns. Human concerns arose from the animal source material under question: in this case, the contradictions and strange mutability of coral flesh. Those evanescent polyps examined by Ellis, Peyssonnel, and others at the seashore and the dried stony relics that Stukeley marveled at in his study formed two sides of the same coin, as coral is inherently dualistic, producing dizzying jumps in scale.

But coral has been unstable on a much deeper timescale, too.

Biologists have long puzzled at reef gaps in the fossil record: the seeming absence of corals and other reef builders in geology's stratigraphic archive for long stretches of time, especially in association with mass extinction events marked by a steep increase in atmospheric carbon dioxide, which tends to spell disaster for marine life. For example, coral biomineralization—the creation of skeletons, and thus, for the scleractinians, reefs—becomes costly in a world of heightened carbon dioxide, which acidifies ocean waters. To this day, individual coral polyps appear to regulate whether to join together as reef-building communities or to sever the bonds of the group based on environmental context. Recent experiments suggest that when surrounding waters acidify past a certain level, polyps may leave their communal form and transmute into solitary, free-swimming, soft-bodied individuals, given how taxing it becomes to deposit a skeleton and maintain connective tissue in an acid world. One study suggests that polyps release enzymes that destroy only the cells joining the animals to the coenosarc—the tissue uniting the community—while leaving intact the cells of the individual polyps. The study's authors describe this release of individuals from the communal bridge as "highly controlled" by the animals themselves. So-called polyp bailout may be reversible, as well; in certain cases, solitary polyps can regroup either when circumstances improve or after polyps swim to a location with better conditions, at which point individual polyps resettle, build a skeleton, and develop a community anew. Some scientists now suspect a plastic ability to alternate between communal and solitary states, and between skeletal and askeletal forms, may have ferried these animals through periods of ocean acidification in deep time. But this very adaptability made corals moving targets to paleontologists, since the soft bodies of unmoored polyps rarely fossilize.[56]

Although the instability of corals has confounded their study by scientists for centuries, the actions of the coral polyps themselves likely paved the way for these animals' survival into the eighteenth century and beyond. Their ability to pivot, both to join as communities and go rogue, let corals *become* the case study for Enlightenment-era naturalists as they sought to resolve some of the central paradoxes of the animal-vegetable boundary. And yet, these animals left a historical record only selectively, and at turns: one of their chief survival strategies in times of environmental disruption has led them to *not* fossilize and therefore *not* leave traces of their activity. Corals made history—they changed over time—in a manner that erased evidence of those changes. The geologist George D. Stanley Jr. writes: "We have come to grasp the distinct probability that diverse corals inhabiting colorful reefs today may, in fact, have had several different kinds of soft-bodied ancestors—ephemeral ancestors, capable at critical points in earth history, of transforming back and forth between calcified and soft-bodied forms. . . . These kinds of relationships create multifarious evolutionary lineages that may be exceedingly difficult if not impossible to unravel and trace through geologic time. Such possibilities place real hurdles before our goal of attaining a coherent and accurate portrayal of the evolution of the group as a whole." Modern coral science is an inheritor of the Enlightenment for its grappling with animal uncertainty, with ambiguity—with doubt—despite the sophisticated technologies, genetic classifications, and monitoring systems at its disposal.[57]

And it is precisely this sense of the unfinished, the ongoing animal history, the *not quite*–ness of coral that could hold clues for their future amid our current global environmental crisis.

CHAPTER TWO

Getting Back

Aboriginal and Torres Strait Islander readers are advised that the following chapter contains the names of deceased persons.

FOR THE LOVE OF CORAL, I found myself ascending a 31-degree incline on an unpaved road in an Australian rainforest. Just before the climb began, a cassowary—the world's most dangerous bird, and one of a diminished 1,500 or so that remain in Australia—watched our car from beside the road, as if to say, "Proceed, if you dare." Nothing can quite prepare a person for a cassowary (figure 2.1 and plate 2). "His Head resembles that of a warrior," one newspaper reported in its account of a live specimen on display in England in 1790. The human-sized bird I saw, with its cobalt blue neck, an imposing keratin crest adorning its head like a Roman galea, and knife-length hind claws that could gut a man like a coral trout, made that a fitting description. It receded into the forest moments later, becoming just another shadow. The encounter had a ring of the mythic. As good a bird omen as any.[1]

I was on the Bloomfield Track, a gravel road that runs parallel with the Great Barrier Reef in tropical Far North Queensland, the northeastern tip of Australia. Most car rental shops in the area strictly forbid taking their vehicles on the track. The road is notoriously impassable in wet conditions, not only for its frightening gradient but also because of rivers that intercept the track over and over like stitching. Those who try their luck on it during the dry season, as

Figure 2.1. A southern cassowary (*Casuarius casuarius johnsonii*) in Far North Queensland. This photograph was taken at a nearby zoo; the image taken of a "wild" cassowary turned out illegibly blurry due to the author's sheer adrenaline. (Photograph by William Robles)

we did, still must drive amphibiously through smaller versions of those waters. (Technically, you should measure the depth of the water before each pass, but the prospect of crocs lying in wait made us take the "drive and pray" approach instead.) Pulitzer-winning journalist and adventurer Tony Horwitz refused to take the road when retracing Cook's voyages for his book *Blue Latitudes: Boldly Going Where Captain Cook Has Gone Before,* writing that even "during dry months, it is a slippery, spine-shuddering track, so filled with craters and fallen trees that traveling it resembles a land version of navigating the nearby reef." Meanwhile, a tourism website for the area proudly boasts of "a true 4WD track with unimproved creek crossings, steep climbs and plenty of opportunity to get stuck if it is wet!"[2]

This postcolonial road is also infamous for the protests it sparked in 1983 and 1984. Conservationists erected barricades, buried themselves in the red Queensland earth, and chained their bodies to trees in attempts to halt bulldozers that had arrived to build the road, which would carve a path through the

Daintree Rainforest, part of Australia's Wet Tropics. This ecosystem harbors a startling percentage of Australia's overall biodiversity: an estimated third of the continent's marsupial species can be found here, nearly half of the bird species, and more than half of the butterflies, ferns, and bats, even though the region covers a mere 0.2 percent of Australia—hence the protests, and hence the area's designation as a World Heritage Site several years later, in 1988. (To the chagrin of environmental protestors, however, local Aboriginal communities tended to *support* the road since it would provide easier access to family members who had been dispersed throughout the region by the forced removals of the colonial mission system. History matters.) The track's dramatic incline, which tempered its environmental impact by reducing the number of switchbacks required and therefore the amount of forest destroyed, was the compromise. The ascent was so steep that I felt my eyes fall into my brain, my heart turn scleractinian.[3]

This rocky road was my access point from Cape Tribulation to Weary Bay—all these doleful place-names from Captain Cook, and all because of some damn coral. Corals had beached Cook's *Endeavour* along this shoreline for weeks in 1770, and they brought me there 250 years later. My husband, William, and I were now bound to locate the aptly named Endeavour Reef that almost took Cook down, hopeful for a near (but hopefully not total) reenactment of his fateful journey, if the trade winds would cooperate. Having waded through abstruse eighteenth-century treatises on polyps, inspected boxes of Linnaeus's own personal coral specimens, opened the many hidden compartments of Joseph Banks's writing desk from the voyage, and strained my eyes deciphering countless hastily scrawled letters, I still had questions about all manner of experiential knowledge that couldn't be found in historical documents or even historical artifacts. Questions about the context not laid down on paper, the sounds and smells and subjective sensations that defined the practice of early natural history. Questions about what it was like to try to behold a polyp as an animal, to see coral in the sea, to fathom or not fathom the scale of a reef, to be a polyp—could one even begin to consider what it was like to be a polyp? Perhaps not, but a human animal—on that front, I had some practice. Above all, I wanted to know what the corals of the eighteenth century might have to say, in so many words, to the corals and the naturalists of now.[4]

As I worked to plan our trip months beforehand, I received a common response from charters and fishers of the region: "You're gonna need a bigger boat." Fortunately, we found the right type of vessel for navigating the protected

Great Barrier Reef Marine Park in the remote town of Bloomfield, population 228. Bloomfield was the kind of place where you thought you felt something crawling across your skin in the night—and it actually, always, was. The kind of place where the stars, in their antipodal patchwork, filled the evening sky to a density you never dreamed possible and made you want to fall supine to the ground—but you knew you'd be envenomed by something unknown to science if you did. The kind of place where the murderous screams of a woman roused you from sleep at 2 a.m.—even though you had been warned in advance not to fear, for it was only a bird. The kind of place where the sole source of sustenance for many miles, a quaint café whose water carafes boasted hand-painted scenes of local animals, was staffed by your ship captain, who was also your lodging host—and who appeared from behind the coffee counter stating, dryly, to your look of surprise, "I wear many hats around here."

The morning of our adventure at sea, we met this captain, our captain, John Corbett, and the ship's first mate, Shane Marks, at the mouth of the Bloomfield River at dawn. Flying foxes flapped overhead, bound home after a night of feasting on nectar. Shane looked at our car and quietly asked John, "They took that on the Bloomfield Track? Is it even an SUV?" (It was. Kind of.) First order of business was actually getting to the boat. This entailed boarding an orange dinghy, deliciously named the *R.I.B. Tender,* which would ferry us to the main vessel. John brought the dinghy to the water's edge, disembarked, and held it awkwardly behind him with one hand, as if creating a barrier—"because of the crocs," he said. Given his droll sense of humor, I assumed this was a joke. It wasn't. Later that day, John and Shane recounted how a massive saltwater crocodile had ambushed and killed Shane's dog right at our entry point.

Once we made it safely from dinghy to ship, we rode forty-five minutes and roughly eighteen miles out to sea as the sun climbed overhead. When we arrived without incident just shy of the coordinates where the *Endeavour* hit the reef in 1770, I was surprised to see that I could, in fact, see it. The reef was not quite the invisible underwater kraken ready to take unsuspecting vessels that I had imagined. A ship's-eye view concealed most of the busy animal world beneath the surface, yet even from a safe distance, we watched the waves faintly dance on the reef's tallest sections. Corals dominated the underworld of the ocean, but they left subtle clues on the other side, too, like icebergs of the tropics. Cook hit the reef at night, despite the bright glow of the moon; John suggested he might have had better luck in the day. When Cook's crew had finally repaired the ship and almost hit a portion of the Great Barrier Reef *again* two months later in August 1770, Cook wrote a torrent of words in his journal explaining that the reefs off Australia's coast were of a sort "scarcely known in Europe . . . a wall of Coral Rock rising all most perpendicular out of the unfathomable Ocean, always overflown at high-water generally 7 or 8 feet and dry in places at low-water; the large waves of the vast Ocean meeting with so sudden a resistance make a most terrible surf breaking mountains high."[5]

We declined approaching any closer, not wanting to sink our own ship on this coral wall. Several digital screens in our boat revealed a steeply ascending ocean floor beneath us, with jagged peaks of coral data shooting like an over-caffeinated heartbeat on a sounding visualization in front of me. As I observed the reef, the site of historic impact, John got straight to the point and asked: "Did you find what you were looking for?" In some sense, I had been looking for a meaningful nothing. Instead, I found something.

Cook hit Endeavour Reef on its southern side, with dangerously churning waters; we navigated around the reef to its slightly calmer northern face for some snorkeling in lieu of an eighteenth-century diving bell. Shane served as our underwater guide while John kept track of everyone from the dinghy. Both warned us as we donned wetsuits and flippers the coloring of poison dart frogs: "You won't hurt the coral, but the coral will hurt you: if you get too close to the reef, use your fins and not your hands as a barrier." In their minds, coral was not the fragile damsel that news reports and contemporary conservation discourse can make it out to be. The reef was potentially malicious, as it had been to Cook, capable of stinging and holding its own, of telling other beings to back off. That sentiment differed sharply from what I knew of other regions, such as the Caribbean conviction that corals should not be touched at all since contact would harm the reef. Perhaps it fit with the Australian mindset that everything was out to kill you.

Once we were submerged in the cold blue ocean, finally privy to the history-making reef, there was no calmly contemplating a single polyp. No quietly beholding the individual animals who together built this grand structure along with their ancestors. There were only multiplying outlines of staghorn and shelf coral, the chaos of feeling one's perception jolted by the waves, and a sense of utter disorientation. (And, it should also be noted, I had far greater swimming proficiency than many early modern Europeans.) It felt inconceivable to *really* know and feel, viscerally, that thousands upon thousands of tiny animals made these intricate structures, eating, secreting, eating, secreting over centuries past. I lacked what William Watson of the Royal Society, in reporting Peyssonnel's coral discoveries, called "that tranquillity of mind, which a just observer should be always in possession of"—and which often evaded even coral fishers in the eighteenth century from their small and vulnerable boats. The next day, I would pick up chunks of coral that had washed ashore at Weary Bay, a bay that made Cook weary since it was too shallow to serve as a safe point for the wounded *Endeavour.* Feeling some sympathy for poor, maligned William Stukeley and his botched attempts to make sense of coral ruins, I'd try to picture a small polyp waving out from each calcified pocket along the surface of the specimens in my hands. But the creatures slipped away from me in situ. Many of the great contradictions, the crises of scale, that these animals posed in the Enlightenment era hold true centuries later: coral as many but one. Soft yet solid. Hard to see, but impossible to miss. An animal still routinely mistaken for a plant. A playful and singular character in the study, in the case of Henry Baker's freshwater polyp, but inscrutable as a subject in context. "Little, poor, helpless, jelly-like animals," according to James Parsons, but producers and destroyers of worlds, all the same.[6]

Another dualism takes on special urgency for this animal enigma's modern condition: coral as imperiling yet imperiled. Despite everything that I could not see below the waterline, I could indeed sense that the same reef that was strong enough to derail Cook had experienced bleaching and degradation (figure 2.2 and plate 3). Corals turn ghostly white when changes in water conditions, such as warming temperatures and ocean acidification, prompt them to expel the symbiotic zooxanthellae that live in their tissues and give the corals color and nutrition. Even though bleaching is not always the death knell of coral, and some do recover, many corals will eventually die without their internal powerhouse partners. Rates of bleaching in the Great Barrier Reef and elsewhere have reached catastrophic levels. It has led biologist after biologist to weep into their facemasks underwater when confronted with the level of damage.[7]

The northern portion of the Great Barrier Reef, where I was, had been particularly devastated by bleaching. There is also a periodicity to the destruction in Australia. The Great Barrier Reef endures what scientists call mass bleaching events, with two major ones—in 2020 and 2022—having occurred between my visit there and the writing of this book. As we surveyed Endeavour Reef, pale exoskeletons of this white blight stood out sharply underwater. More noticeably, algae had overtaken many of the dead corals, covering large

Figure 2.2. The author over Endeavour Reef, with some bleached corals, some in good health. (Photograph by William Robles)

portions of the reefscape with a dull gray slime (figure 2.3). Algae can sustain its own forms of biodiversity, but it tends to eliminate potential growing surfaces for future corals and indicates an unhealthy reef environment if it supplants coral too quickly. To make matters worse, crown-of-thorns starfish in the region consume coral and proliferate when sea temperatures rise. And a recent cyclone at Endeavour Reef had also macerated portions of the structure, scattering coral remnants across the seafloor like bones in a submarine midden. Corals face threats at interlocking scales: from individual animals and people, from discrete weather events, and from global climate systems.[8]

All the same, bursts of color punctuated the reef in the form of healthy corals, while schools of fish, giant clams, nudibranchs, and a wandering shark reminded us that even degraded reefs could support lively animal communities (figure 2.4). I wasn't sure what to think, especially since I lacked any comparison, this being my first visit to this site. Perhaps our baseline had simply shifted. Shifting baseline syndrome is one of the perpetual concerns of ecologists and environmental historians: when environments lose biodiversity, so

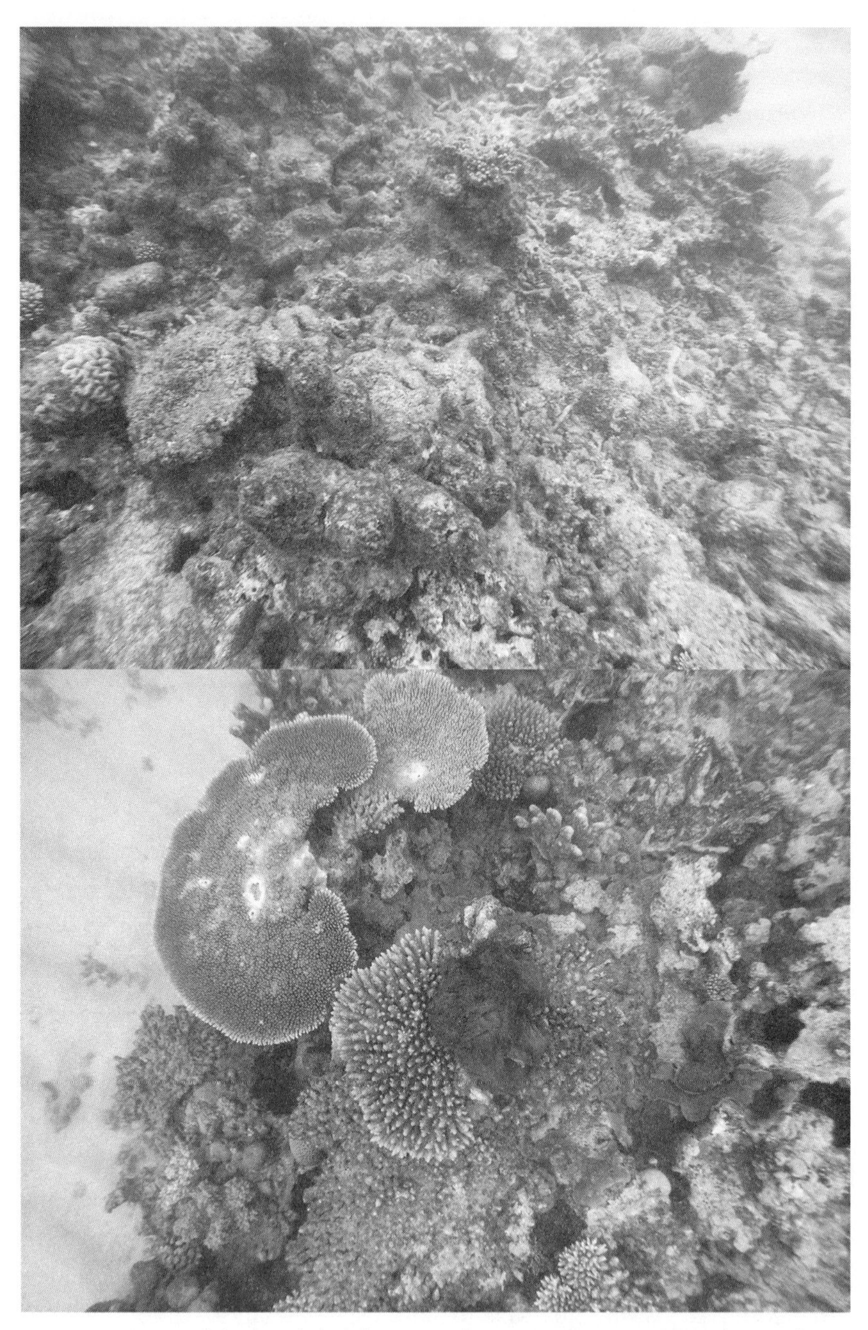

Figure 2.3. *Top:* Coral death and algal slime on Endeavour Reef. *Bottom:* Macroalgae settling on corals in Endeavour Reef. (Photographs by Whitney Barlow Robles and William Robles)

Figure 2.4. Fish swim through Endeavour Reef. (Photograph by Whitney Barlow Robles)

the idea goes, it becomes all too easy to accept a new normal as the standard or natural condition without realizing what came before.[9]

Yet a baseline is still a human-made concept, one with an implicit preference for no humans at all. And everything in Australia's seascape seemed to be shifting all the time. After our journey through the reef and a tidepool pit-stop where we squirted sea cucumbers at each other, the four of us ate shrimp sandwiches prepared by John at one of the Hope Islands, named so by Cook because he was "always in hopes of being able to reach these Islands." John and Shane informed us that the entire island changed its orientation after the cyclone, spinning across the ocean like the needle of a compass. This ever evolving landmass, perpetually outdating maps, was a coral cay. Such islands form from reefs so tall that they approach the water's surface, becoming bona fide land with the aid of sediment-carrying currents and the blessings of shit: parrotfish use their fracture-defying mouths to eat coral and excrete it as island sand, while the nitrate in bird droppings nourishes the island's plant growth. (And lest I sound crass, I gesture toward the endnote and what I brand the shit turn in environmental history.) A low-lying coral cay like this island is extremely vulnerable to climate change. Rising sea levels could swallow it completely. But the wandering, liminal nature of the terrain also made it harder to see where nature's resilience ended and the environmental degradation began. We were careful to avoid leaving any seeds from our lunches, not wanting to introduce new plant species that might overtake the tiny ecosystem, yet another reminder of the outsized consequences of small actions, human or coral.[10]

Before long, we had to get back. Cook had thought he was done for the evening; in June 1770, the crew was slowly making its way back to England, charting the eastern coast of Australia in the process and eventually claiming it for the crown. The night of the impact on Endeavour Reef, Cook settled himself into bed as the ship headed away from shore, unaware they were ensconced in a great lagoon rimmed by coral. John, with the help of a cartoonish map laminated to the surface of the only table in our ship, showed us the network of reefs Cook must have accidentally and narrowly dodged before hitting Endeavour Reef. Cook would only begin to fathom it later as he observed the cascading chain of coral hazards from a high hilltop with a mixture of wonder and horror. Having unknowingly maneuvered this "insane Labyrinth" before striking Endeavour Reef just before midnight, Cook's dumb luck ran out. His minotaur awaited. Only this boss would be fought in pajamas.[11]

I, too, thought my research was done for the day. I had seen what I didn't know I was looking for, and I had pictures to prove it. But those trade winds that behaved so well in the morning had picked up during our jaunt on the island. The much longer boat ride to shore proved more harrowing than homecoming, as our ship jumped porpoise-like through the air with each crest, over and over, for a good hour. I hadn't been looking for this, either, but it gave me a renewed appreciation for how fortunate the broken *Endeavour* and her crew were to survive. The coral that gouged Cook's ship was part kraken after all, stranding them in an environment that could change at a moment's notice and dunk them into a medium hostile to human physiology, even as the lush mountains of the coast remained tauntingly visible the whole time. Unbeknown to the voyagers, they had also just barely avoided stinger season, the yearly reign of the deadly Irukandji and box jellyfish along Australia's shores. William began to vomit out the window, incurring banishment to the open-air back of the ship. I checked every few minutes to make sure the waves hadn't tossed him into the lair of the coral.

Once we arrived safely back on shore, we made our own journey to Cook-

Figure 2.5. Retrieval of a coral-encrusted cannon from the *Endeavour* on February 11, 1969. (Photograph by Arthur Keith Miller of Cooktown)

town, an hour north by road, and to the Cooktown Museum, housed in a nineteenth-century convent by the sea. When Cook and his crew finally managed to get the *Endeavour* to shore for repairs, they took it through the mouth of Waalumbaal Birri, also called Endeavour River henceforth, at present-day Cooktown. Corals are the reason Cooktown, as Cooktown, exists. Upon entering the museum, one's attention settles instantly on a massive *Endeavour* anchor and cannon recovered from the ocean floor. Cook's crew desperately threw its cannons, iron and stone ballast, and several other ponderous objects overboard to lighten the ship once they hit the reef. The implements lay covered in cementlike sheets of coral until an expedition of Philadelphia's Academy of Natural Sciences rediscovered them using a magnetometer in 1969 (figure 2.5). Some of the items were only freed from the obstruction of coral by underwater explosives. Coral encrustations that had preserved these metal Cook relics for two centuries now sat on the museum floor beside the anchor. Coral had a way of making things stick together, of uniting human history and animal history in a single object lesson.

Cook was a colonizer. Mere weeks after repairing the *Endeavour*, he claimed

Australia's eastern coast for the crown at what he called Possession Island. I was a settler-historian not unproblematically trailing his path. But corals were colonizers, in their own way, too. Polyps were indifferent to the human history, imperial dreams, and misplaced nostalgia infusing these expedition artifacts: the sunken guns were merely good surfaces to fasten to, things to settle on, occasions for more reefs. Those chunks of coral on the ground next to Cook's towering anchor raised more questions than they answered. Who was colonizing whom? Who was anchoring whom? Who had collected whom, who was using whom, and whose history was at stake here?

Across from the anchor and cannon exhibit, two contrasting perspectives of Cook's arrival in the region faced off. One placard featured journal excerpts from the *Endeavour*'s Europeans and another the words of Eric Deeral, an elder of the Gamay Warra clan of the Guugu Yimithirr, the Aboriginal group that encountered Cook at this site in 1770. The quotes from Cook, Banks, and other voyagers stressed the danger of the coral encounter, conveyed excitement at finding massive sea turtles as a food source, and puzzled at an unfamiliar animal resembling a greyhound in form and a rabbit in locomotion—a kangaroo, which they eventually caught and ate, too. Eric Deeral's account described an initially peaceful exchange of fish and beads between Cook's men and the Guugu Yimithirr. But when the latter inspected the *Endeavour* and noticed the cache of sea turtles amassed by the strangers, animals taken without permission, the voyagers refused to part with even one. In Deeral's words, "something had gone wrong": Cook's campsite set on fire, a Guugu Yimithirr man shot. At last, a tenuous peace was made. One empty turtle shell accompanied the corals on the exhibit floor.[12]

Millions of brainless polyps, and not long thereafter the turtles, had precipitated this momentous meeting between European and Indigenous communities. Human histories of science, cultural encounter, and colonialism have always been intimately interwoven with animal history. June 2020 marked 250 years since Cook's arrival in Far North Queensland after his tussle with the reef. But for both settler and Aboriginal communities in the region, that history has never been simply something of long ago; time feels recursive here, the past ever present, the shadow of Cook's sails always waving over the shore, his name on the tip of the tongue. There is room for creation amid the residue and destruction of this history. The journals produced during Cook's expedition, for example, served as key evidence for the Guugu Yimithirr of Hopevale to become some of the first Aboriginal people on mainland Australia to secure

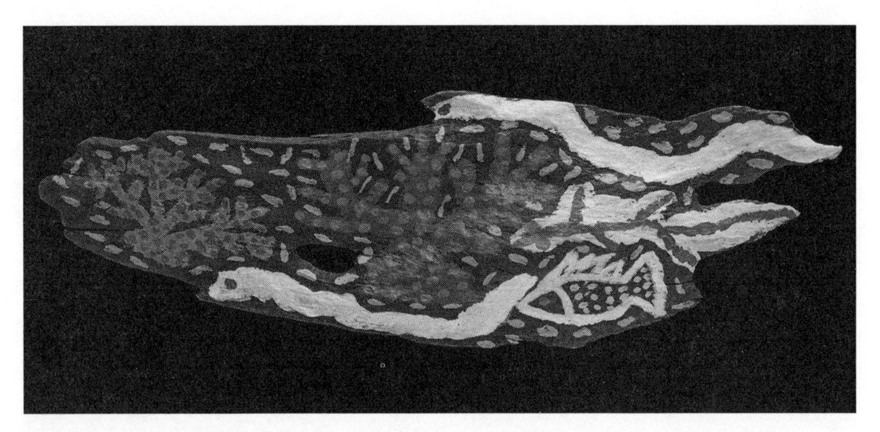

Figure 2.6. Josie Olbar, *Dulngku* (coral reef), 2019. An acrylic painting on driftwood featuring coral, sea snakes, and a fish. (National Museum of Australia [MA103108141]; photograph by Jason McCarthy; © Josie Olbar/Copyright Agency; licensed by Artists Rights Society [ARS], New York, 2022)

legal title to their ancestral lands, in 1997. And Cooktown hosts an annual reenactment of Cook's visit to the region, an ever updating event that contends with the complexity of both the violence and reconciliation that ensued from the encounter.

And then there is art. On the occasion of the 250th anniversary, the National Museum of Australia displayed a collection of driftwood artworks made by Aboriginal artists, who collectively identify as Bama, based near Bloomfield. "The history of Australia would have been very different if the *Endeavour* had not been freed from the reef," notes the museum. Bloomfield sits near the community of Wujal Wujal, home to the Kuku Yalanji, Kuku Nyungul, and Jalunji peoples as well as the Bana Yirriji Artist Group. According to a statement from the group in 2019, these artworks show "the other side of the Captain Cook story." They feature imagistic stories painted on driftwood collected at Weary Bay, a medium chosen to evoke the wooden chunks of the *Endeavour* and broken coral fragments Bama ancestors would have found washed ashore in 1770 (figure 2.6 and plate 4). Coral features prominently on many of the pieces, which act as time machines to help the artists envision the sights their forebears must have beheld in the wake of Cook's unplanned arrival, even as the artworks also look ahead to confront the distressing bleaching besetting the reef, a key source of food for these communities. Josie Olbar, a Kuku Yalanji/Jalunji artist, said of her vibrant coral painting: "Cook found it

Figure 2.7. Gertie Deeral, *Dangerous Reef,* 2019. A lightbox made from plywood and medium density fibreboard featuring an acrylic painting. (National Museum of Australia [MA99250097]; photograph by Jason McCarthy; © Gertie Deeral)

hard sailing through the Great Barrier Reef. His ship, the *Endeavour,* struck the reef not far from here. Our dulngku [reef] is dangerous to people who don't know their way around it." Gertie Deeral, a leader and elder of the Dingaal clan, also paints in response to Cook's arrival in the region. Her work *Dangerous Reef* (2019) illustrates the weight of animal history as an equal and opposing force to human history, albeit one that is submerged (figure 2.7 and plate 5). Coral polyps shine like blinding little suns beneath the *Endeavour* as it runs aground on the reef. The crew jettisons all it can overboard. But Cook, of course, did not collide with coral alone: as shown by Deeral's relationally minded image, he encroached on an interwoven community of beings, the connections among corals, rays, turtles, fish, and many other species, including humans, forever changed. In fact, this book's own laser-like focus on four creatures—extracting particular animals as the guiding superstructure for its narrative like putting specimens in boxes—is kin to a colonial ethos. "Bama," says Deeral, "know the reef is dangerous and that you need to know it very well."[13]

As for the *Endeavour* itself, it might yet find a delectable fate. A wreck thought to be the famed ship, which was later renamed and reused as the *Lord Sandwich,* now lies off the coast of Rhode Island. The British sank it intentionally in 1778 as part of a blockade to fend off French aid during the American Revolution. Whatever the ship's true identity—an ongoing matter of dispute between American and Australian institutions—its remnants are today being chewed into oblivion by wood-hungry gribbles and shipworm.[14]

We thought that the Bloomfield Track should be smoother sailing downhill. A much longer inland route to the airport in Cairns would have allowed us to bypass it completely. But the purest of motivations led us to risk the treacherous road again: ice cream. Before we had begun our initial ascent to Bloomfield days earlier, we steeled ourselves at an ice cream stand in the jungle that made our mouths burst with flavors we had never experienced—the nuttiness and coffee tinge of wattleseed, the bright tang of lemon myrtle. It was only once I returned to America that I would read on the tourism website: "Descending is often more dangerous. Select low gear beforehand, keep speed down using engine braking where possible. Never rely on brakes alone. If you start to slide accelerate to straighten the vehicle—braking makes the slide worse."[15]

Sure enough, the descent downhill caused our car to critically overheat. The notion of cellular service in that area was laughable. William grabbed the twelve-pack of water bottles we had stowed for an emergency and opened the hood. As he poured our spare supply over the steaming engine to cool it into submission, I was skeptical and afraid: I had heard of flooding the engine, and this seemed like a hell of a good way to do it. Fortunately, William's mechanical knowhow surpassed mine, and we were on our way again. We arrived at the legendary mountainside ice cream stand only to find the shop closed for the day. A jeep ahead of us, apparently with the same idea, whipped around in dejection. "Some things are meant to be had just once in a life," William tried to console me. "Makes them special."

On the highway back to Cairns, another driver honked and flailed his arms wildly at us. We thought he must be celebrating our bravery to face the Bloomfield Track, surely self-evident from the coat of mud on our forced-amphibious vehicle. He was probably signaling about a flat tire that we didn't discover until we stepped out of the car at the rental shop. From Cairns, we were bound for Sydney and then home. During the transpacific flight to

America, I fell asleep. My dreams steered me through steep hillsides, soaked in the honeyed light of the golden hour, covered in hundreds of kangaroos lying down for rest. When I awoke to the dim blue glow of the cabin, I felt a pang at the thought of return.

Once home, however, I wasn't done needing to get back. I now had an archive of photographs and video footage of the reef to make sense of, to try, amid the comfort of dryness and climate control, to recall my experience in the wet and dirty field and the animals that had so eluded me there. I met with a coral biologist named Aaron Hartmann at Harvard University, who generously offered to look through my footage and help assess the state of Endeavour Reef. We sat down next to each other at a conference table and began scrolling.

Aaron agreed that the reef had experienced some coral mortality and was what he termed "relatively degraded." But putting our finger on how and why was far from straightforward. Very broadly, for a reef to be considered healthy, Aaron said about 50 percent of a given visual snapshot should contain live corals—at least for Aaron's own baseline of Caribbean reefs, his research specialty. To calculate that percentage, Aaron made a rectangular frame with his fingers to break up each photo into smaller bits. We did note many live corals, including perfectly healthy and naturally white *Acropora,* some of which I had mistakenly assumed were bleached. Aaron also noted the presence of beneficial pink coralline algae. In addition to cementing reef structures together, this algae harbors bacteria that can send chemical signals to free-floating coral larvae letting them know it's a good place to settle, to call home.

But we concluded that for many of my photographs, only around a quarter of what we could see was live coral. And much of the dead coral, although it was the type you'd see on any reef, was so near the top of the structure that it must have died only a couple of decades ago rather than centuries earlier. Moreover, a good portion of the visible algae was not a welcoming party but rather the sort that threatened coral. "Have you heard of shifting baselines?" Aaron asked, his eyes still fixed on the screen as we continued to wade through the photographs. I told him I had. "It's even a problem for scientists who have been doing this forever, and who are really quantitative," he said, eyes unmoved.

How do you get back when all you can do is move inexorably forward?

In my own process of trying and floundering to get back, to understand how we got here through the lens of the eighteenth-century past, I could see that many of the problems now facing biologists both mirror the queries of Enlightenment-era naturalists and were, in fact, produced by them. At least in the realm of Western science, the eighteenth century was the moment when corals finally became animals. It's thus when many of their contradictions intensified, and when some first began. The circles of coral can't be squared, won't relent. Whether these creatures are endangered or dangerous, small or vast, plant-like animals or animals that contain plants, the paradoxes hold as paradoxes, an unyielding doubleness and multiplicity as intrinsic to the very being of coral as a crest is to a cassowary.

Early naturalists ranked polyps at rock bottom among the animals. Yet the rocky bottoms polyps generated made the likes of Captain Cook and Joseph Banks quake with fear. Coral polyps remain exceedingly small animals, some measuring just a few millimeters in length, but in a reversal of the Chain of Being they enact powerful effects on our globe. They maintain and protect coastlines and the people who inhabit them. As ecosystem engineers, they build homes for the most diverse animal communities on the planet, in turn sustaining fisheries and feeding millions. And they contain within their deposited skeletons, in the labor-made-matter of their ancestors, a stone archive of oceans past that scientists now use as a record to understand how climate has changed throughout the deep history of our globe. Conservationists and biologists are desperately scrambling to stem the tide of coral death. Some are working on local, intimate, embedded solutions, such as growing corals in labs and aquaria, or even in marine nurseries where they attach corals back to reefs and then lovingly tend the new arrivals in an act of animal gardening. Others insist on tackling carbon emissions and ocean acidification writ large, prioritizing global, total-system change.[16]

What room, if any, does human-driven climate change leave for the resistance of animals? For the animals whose power threatened the empires of early modernity? Semi-satirical "obituaries" of the Great Barrier Reef have recently proliferated in digital and print media. "In lieu of flowers," concludes one such piece, "donations can be made to Ocean Ark Alliance." But such statements belie the complexity of coral's modern quandary, and above all may underestimate corals, which have, as with so many other binaries, found ways to muddy the one between the living and the dead. These animals, both moribund yet theoretically immortal, have been here far longer than we have, and they may outlive humanity at last. Precisely what coral *is* is not even stable, given the ability of polyps to change between solitary and grouped states, to shirk their symbionts and re-welcome them later with open arms, to shed their skeletons to wait out eras of turbulence.[17]

Corals' adaptability to ocean acidification and other environmental stressors could help them weather our current ecological tempest, even though the Great Barrier Reef and reefs the world over will no doubt be forever altered by anthropogenic impact. Make no mistake: humans are culpable, and we must be co-conspirators in creating a future for corals. In *Coral Whisperers,* a compelling study of the emotionally charged work of modern coral conservation, Irus Braverman describes how coral scientists are caught in a loop of "despair and hope" when faced with the anthozoan outlook, inhabiting a kind of "split reality." How corals might endure the daggers of climate change, runaway carbon emissions, rising temperatures and sea levels, pollution, and ocean acidification—whether through toggling between individual and aggregate forms, moving to cooler or deeper waters, allying themselves with more heat-resistant algal symbionts, or perhaps some other creative solution that only corals will dream up—is still riddled with uncertainties. Their persistence will almost certainly require symbiosis with our species, such as in the form of human-assisted migration or human-assisted evolution, though these roads ahead are hotly debated in the coral community. Scientists who are working to study the mechanics of how corals might survive the next century must pepper their findings with caveats: "we still do not know how widespread this behavior is in anthozoans"; "we do not know . . . whether this behavior is simply an artefact of laboratory culturing"; "we do not know."[18]

Some eighteenth-century thinkers located the source of polyps' wonderful powers in their vulnerability. James Parsons wrote: "If we only observe their extreme Tenderness, which exposes them to be wounded, nay, torn to Pieces, by any hard Body . . . we may easily see the providential Reason for placing Organizations everywhere for their Restoration and further Propagation; for perhaps, there is no other Animal of so tender a Texture, and consequently so easily destroy'd." Perhaps the inner unresolvability of corals—their ultimate unpredictability and unknowability—should indeed inform our response to coral's current plight. Dynasties of coral have done something that humans have not: survived mass extinction events over the course of millennia, though always barely, always in changed form, and usually with gaps in reef formation extending millions of years, at least from what we can read in the rocks. We are currently midstream in a probable sixth mass extinction event, with rates of loss exceeding a "normal," "background" rate by hundreds or thousands of times.[19]

As yet, there's no telling which species, including ours, will make it out alive. But panic and hope for coral need not be mutually exclusive. Neither must love and fear. Against this sea change in the history of life, we might cultivate Captain Cook's dread when we behold these besieged beings: a nearly biblical, sublime fear that mingles wonder with an uncanny awareness that these waters were not made for us and may before long swallow us whole in a bed of sea whip and staghorn.

RATTLESNAKES

For he seemed to me again like a king,
Like a king in exile, uncrowned in the underworld,
Now due to be crowned again.

—*D. H. Lawrence, "Snake"*

CHAPTER THREE

The Lost Serpent

CALL HER 1.11.1.A. Sometimes catalog numbers are the only names we have. She has been folded accordion-style in a bottle of liquor for close to three centuries, if not longer. Bleached scales encase her form, evoking what the anatomist Edward Tyson in 1683 called a "coat of armor for their defence," so "curiously contrived" for creatures that "must be always grovelling on the *ground.*" Turn her transparent tomb, heavy with snake and liquid, and surgical slits running the length of her large body grow conspicuous. Through these openings, 1.11.1.a was prepared for the afterlife—and perhaps emptied of her young, had she been pregnant. Naturalists often wrested fully formed neonates from those unusual serpent species, like this one, who gave birth to live offspring while basking communally near their underground lairs. Trace the curves of her form from her head to her curious terminus at the bottom of the bottle, and a segmented appendage appears like an ear of corn, identifying her to even an untrained eye as a rattlesnake (figure 3.1, figure 3.2).[1]

Some would say she shouldn't exist—that she is an impossibility, of a certain breed. Faded labels at the top of her bottle bear the name of Sir Hans Sloane, as do several catalog entries that also assign her a female identity. Sloane, a naturalist and eventual president of London's Royal Society who collected everything from purses made of asbestos to manuscripts on snake charming in Asia, amassed thousands of animal specimens in his day. Now, almost none remain. After Sloane died in 1753, his collections laid the foundation

Figure 3.1. Timber rattlesnake (*Crotalus horridus*) specimen, labeled 1.11.1.a, from Hans Sloane's collection, late seventeenth or early eighteenth century, housed in the Herpetology Collections of the Natural History Museum, London. (Photograph by Whitney Barlow Robles; by permission of the Trustees of the Natural History Museum, London)

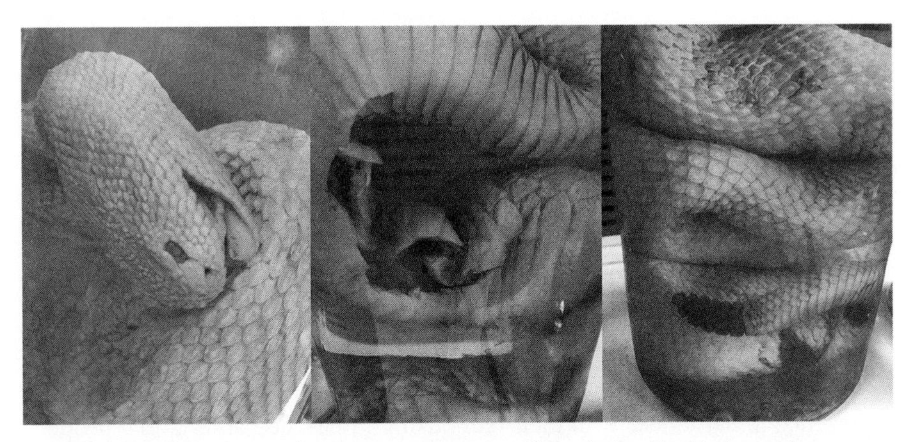

Figure 3.2. Details of Hans Sloane's 1.11.1.a specimen of *Crotalus horridus* (timber rattlesnake). The bleached body suggests that she was displayed for some time, as light exposure can cause natural history specimens to fade, or that the alcohol used to preserve her body robbed it of color. (Herpetology Collections of the Natural History Museum, London; photographs by Whitney Barlow Robles; by permission of the Trustees of the Natural History Museum, London)

for the British Museum. In the early nineteenth century, however, a series of keepers at that institution set out to ritually torch the bulk of his zoological specimens, which they found too grotesque, decayed, and unsystematic for their new age of ordered collections. George Shaw and William Leach were said to have called their yearly blazes "cremations," and in the words of one witness the bonfires wafted "a pungent odour of burning snakes." Animal specimens from Sloane's founding collection such as this bleached serpent, then, have been a white whale for museum staff: elusive but eagerly sought. The reptiles curator at London's Natural History Museum—a former arm of the British Museum that retained most of its scientific holdings—suspected there were no eighteenth-century serpents in the museum's collections at all when I visited its storehouse in search of early modern animals. A Sloane specimen had not even been on our minds until I chanced on his name in their on-site catalog. Some digging in underground storage led us to I.II.I.a, a timber rattlesnake, among a cache of other centuries-old rattlesnake specimens. For all the stench of burning snakes, I.II.I.a had been under the museum's nose the whole time.[2]

Though she may not look it, I.II.I.a is a relic of a contentious struggle over the nature of animal power and, in particular, the power of the uniquely American rattlesnake. She is also an emblem of what happens to knowledge in the wake of annihilation. If coral polyps were the lowest of the low to early naturalists, submerged like reefs at the base of the Chain of Being's animals, snakes who slithered across the ground on a different plane of existence while recalling humanity's original sin did not fare much better. So too, as corals had helped and hindered the people who studied them, rattlesnakes likewise both attracted and repelled the curious human. In 1758, Linnaeus gave timber rattlesnakes the name *Crotalus horridus,* its genus deriving from the Greek *krotalon,* a rattling instrument similar to a castanet, and the species name *horridus* referring to the animal's scaly texture. Rattlesnakes resolutely announced their presence to naturalists with their hair-raising rattles, demanding that people look and listen. They enchanted and charmed, drawing observers close. Of all the creatures native to the Americas, rattlesnakes may have spilled the most ink from the quills of naturalists, physicians, explorers, surveyors, and garden-variety settlers, while also drawing a rich oral record from Native peoples and enslaved Africans. In fact, one traveler in 1749 declined to comment on rattlesnakes, calling them "too well known to need any Description."[3]

Yet alongside this information overload, rattlesnakes also damned efforts to understand them. They did so in ways indebted both to their own cryptic biology and to deep-seated cultural beliefs and fears among humans. In 1792, the Philadelphia naturalist Benjamin Smith Barton looked back at the centuries-long scientific debates about rattlesnakes and concluded in no uncertain terms: "The truth is, that there is no branch of natural history in the investigation of which even men of science have more prominently discovered their ignorance and weakness than in that of the serpents."[4]

Visceral human attitudes toward these animals marked, and indeed still mark, their natural histories with loss. In the eighteenth century, some naturalists wondered if a full scientific accounting of rattlesnakes would best be left unfinished. *Not* knowing rattlesnakes, by not encountering them, could yield physical safety. Meanwhile, Native peoples, bound to rattlesnakes in complex webs of obligation, kept European invaders ignorant of the physical whereabouts of rattlers even as they provided critical information about the animal's natural history to inquiring scientists. Rattlesnakes took on special significance among the New World's dangerous animals not only because they provoked extreme dread and disgust but because they were deeply subterranean beings. Rattlesnakes ruled the very ground beneath settlers' feet—land settlers wished to claim for themselves. As a result, targeted violence has also obscured the natural history of rattlesnakes by wiping out their societies: colonists experimented with annihilating rattlesnakes at a time when the possibility of extinction remained an open question, while gendered violence, committed by scientists and settlers toward female snakes such as I.II.I.a, threw the existence of the species into further precarity given the paramount role of female animals in rattlesnake communities. Timber rattlesnakes, denizens of the eastern half of North America, have since disappeared entirely from Canada. They are almost just a memory in New England, one of the prime early contact zones between Europeans and the species. Rattlesnakes have been both hypervisible and nearly invisible, equal parts powerful and vulnerable.

Above all, rattlesnakes themselves—and not merely the prejudices humans brought to them—evaded the naturalist's gaze in the eighteenth century, just as they continue to do. The specimen I.II.I.a might have been a lost serpent, but in some sense it is the nature of a rattlesnake to get lost. They are master hiders. Darkness and occlusion are to them what water is to fish, wind to birds. Much of a rattlesnake's life unfolds well beyond the bounds of human experience and oversight. Timber rattlers spend most of each year commanding the rocky channels of underground fortresses beyond human purview. Their remarkable social structure, which Euro-American scientists recognized only recently, plays out in shadows as the snakes curl up and consort with their kin in subterranean dens, safe from the frosts of winter. In this hypogeal commons, they communicate with each other and understand the world via pheromones, vibrations, and infrared visions invisible to us. They are, to quote Donna Haraway, "significantly other." Rattlesnakes thus posed a puzzle to early natural history: how to study a creature who did not wish to be found? Especially when some naturalists wished not to find it?[5]

TROUBLE IN PARADISE

Coral reefs made getting to new worlds difficult. Rattlesnakes posed fearsome barriers once colonizers stepped on land. They likewise summoned rich symbolism that could be difficult to disentangle from their physical being. Snakes signified the New World's status as both Eden and hostile wilderness, and they eventually became distinctly American emblems—most famously in Benjamin Franklin's "Join, or Die" print and the now familiar yellow Gadsden flag—at the very same time their bodies procured bounty payments (figure 3.3). Rattlesnakes are, moreover, blatantly semiotic animals, the rattle acting as the creature's own signaling apparatus. In the words of a present-day naturalist, it

Figure 3.3. Benjamin Franklin, "Join, or Die" woodcut, *Pennsylvania Gazette,* Philadelphia, May 9, 1754. The pose of the snake in this famous print mirrors contemporaneous scientific illustrations, which often fit a serpent's prodigious length on the page through looping and scrunching its body into low-amplitude waves. The other dominant strategy for depicting rattlesnakes in natural history visualizations—the coil—proved useful in showcasing an agitated rattlesnake on the Gadsden ("Don't tread on me") flag. (Library of Congress)

may well be "the most novel appendage in the animal kingdom." Given the rich symbolism associated with rattlesnakes and their capacity to speak back, in some limited sense, to humans, perhaps it is unsurprising that most of the many historical studies of rattlesnakes have approached the animals not as flesh-and-blood beings, but as rife metaphors, cultural representations, or mere proxies of human politics. Thomas P. Slaughter reflected this scholarly tendency when writing of early American rattlers: "Snakes were the least of it; but the symbol is stronger than the serpent itself."[6]

And yet—what of the serpent itself? Snake symbolism and the figurative rattlesnake have proven so potent in the historical memory of early America that scholars have too often overlooked the actual creatures, such as I.II.I.a, and their material omnipresence in natural history and everyday life.

As the English established North American colonies during the early seventeenth century, this primordial snake in the primeval American garden proved biblically foreboding and drew special venom from arriving settlers. Francis Higginson, a Puritan minister and early recruit to the Massachusetts

Bay Colony, wrote in his promotional *New England's Plantation* (1630) of the abundant wood available for ship masts in America, arable land conveniently cleared of trees by Natives, air so fresh that one "sup" proved "better then a whole draft of old *Englands* Ale," and a bounty of Indigenous corn awaiting English settlers. Still, devoted to "nothing but the naked truth," Higginson also reported several "discommodities" in paradise: chief among them mosquitoes, frost, and rattlesnakes. "This Countrey," he wrote, "being verie full of Woods and Wildernesses, doth also much abound with Snakes and Serpents of strange colours and huge greatnesse: yea there are some Serpents called Rattle Snakes, that haue Rattles in their Tayles that will not flye from a Man as others will, but will flye vpon him and sting him so mortally, that he will dye within a quarter of an houre after." An English traveler named William Wood similarly reported "what is evill" in New England, deeming rattlesnakes public enemy number one, as "that which is most injurious to the person and life of man."[7]

Most accounts agreed that there was something singular and superlative about the creature. By the eighteenth century, many natural history compendia included rattlesnakes as the first entry among serpents and other reptiles, a signifier of an animal's importance that derived from the structure of medieval bestiaries, which often listed the lion, king of beasts, first. English naturalist Mark Catesby placed the rattlesnake first among his serpents in his lavishly illustrated tome on the flora and fauna of the Carolinas, Florida, and the Bahamas, the culmination of his New World travels in the early part of the century. He called the rattlesnake "the most formidable, being the largest and most terrible of all the rest." The Massachusetts jurist Paul Dudley believed that this fear crossed species lines, for it seemed that "both Men and Beasts are more afraid of them, than of other Snakes." Yet the idea of an Eden overrun with serpents became somewhat easier to swallow when the snakes could, miraculously, alert humans of impending danger. Pehr Kalm, a disciple of Linnaeus who traveled North America from 1748 to 1751, noted that this trait even made the animal vulnerable, for, "because of its warning rattle, the snake betrays its location and can be destroyed quickly." Although most biologists today believe the appendage evolved to ward off would-be predators, to eighteenth-century observers, the rattle's uncannily loud and foreboding sound seemed to be a built-in alarm system crafted by God to protect humans.[8]

Even so, rattlesnakes made the day-to-day work of natural history and exploration risky. They threatened the production of natural knowledge writ large perhaps more than any other creature. Consider the anxieties of the

London merchant-botanist Peter Collinson, the same man to whom William Stukeley penned his diatribe against the animal nature of coral. Collinson wrote to his frequent correspondent John Bartram, a Quaker naturalist and avid rattlesnake observer based in Philadelphia: "It is with pleasure when we read thy Excursions (& wish to bear thee Company) but then it is with concern that we reflect on the Fatigue the[e] undergoes the great risks of thy Health in Heats & Colds but above all the Danger of Rattlesnakes." He confessed that those snakes "would so curb my Ardent Desires to see vegitable Curiosities that I should be afraid to venter in your woods unless on Horseback & so Good a guide as thee art by my side." For Collinson and many others, acquiring knowledge of American nature had to be weighed against the risk of rattlesnakes; the very thought of encountering these animals could blunt curiosity's force. John Bartram and his son William, some of the standout rattlesnake sympathizers of their time, could also be stopped in their search for knowledge by this potent being. William recalled one close brush with a rattlesnake during a botanical expedition with his father that "deterred us from proceeding on our researches for that day." As a precaution against this foiler of scientific study, Pehr Kalm recommended a change of clothing for those working on the ground in American landscapes. Boots wouldn't do, since the long fangs of a rattler could still penetrate them. "It is safest to wear wide sailor's pants," Kalm concluded, "which come down to the shoes. If the rattler strikes the cloth of the pants, it buckles or wrinkles, making the strike ineffectual. Therefore loose pants are safer than boots." Even the daily rhythms of material culture responded to the threat of rattlesnakes.[9]

The palpable influence of these animals extended to other realms of knowledge-making beyond natural history proper. As subterranean creatures camouflaged in the continent's topography, rattlesnakes also circumvented human attempts to understand and control American landscapes through surveying, mapping, and clearing. Surveyors were especially liable to encounter rattlesnakes as they drew boundary lines in the snakes' territory, where they might step mistakenly on an unsuspecting serpent. John Lawson, an English explorer in the Carolinas, claimed that "the most Danger of being bit by these Snakes, is for those that survey Land in *Carolina*." Likewise, William Byrd II, a planter and slaveowner who served as commissioner for the 1728 survey of the disputed boundary between Virginia and North Carolina, blamed rattlesnakes for a months-long delay in the surveying activity of this contested landscape, variously claimed by the Meherrin Indians, settler communities,

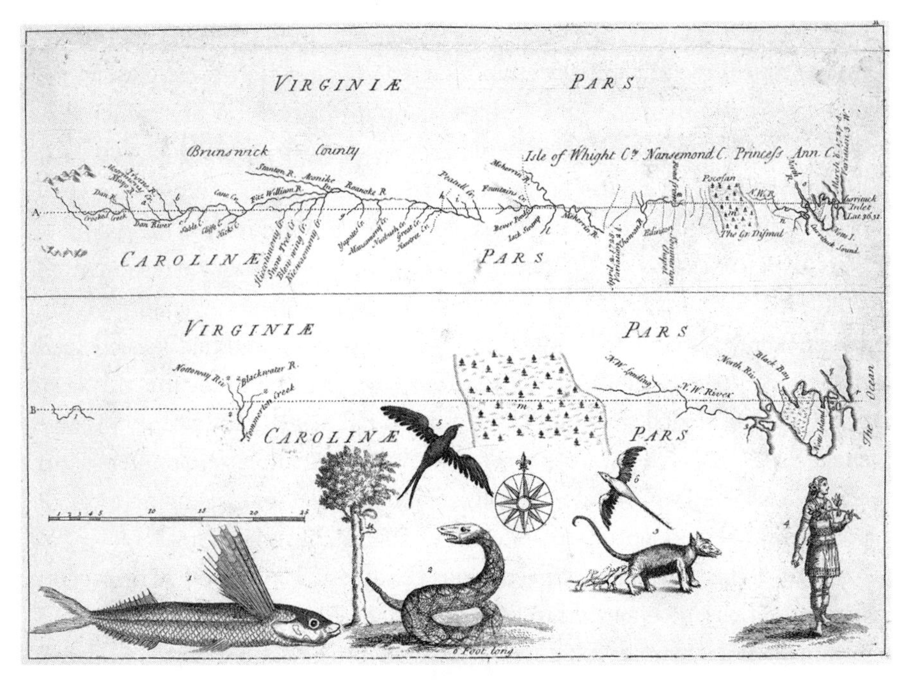

Figure 3.4. An engraving of the Virginia–North Carolina dividing line, surveyed in 1728, featuring a rattlesnake enchanting a squirrel. The engraving was most likely intended to accompany a publication of William Byrd II. (Rawlinson Collection Copperplate c. 29* [strike]; The Bodleian Libraries, University of Oxford)

and snakes. Byrd took the springtime emergence of rattlesnakes from their underground lairs as reason enough to suspend the expedition into the fall. A copperplate of the dividing line, most likely intended for publication by Byrd, prominently featured a rattlesnake on the sketch of the land itself (figure 3.4). Earthbound rattlesnakes impeded settler efforts to understand, and thereby possess, American nature.[10]

In many cases, these practical threats were no doubt heightened by ophidiophobia, or the fear of snakes, one of the most common phobias in the human species. (Some readers with this affliction may have already frantically skipped this chapter and the next—an evasion with tangible consequences for understanding serpents.) Whether primates are evolutionarily predisposed to fear snakes is a subject of much scientific discussion and veers into a thorny thicket of determinism; nevertheless, a belief in innate loathing of snakes surfaced in many Enlightenment-era accounts. Collinson insisted on a "Sort of Natural aversion in Human Nature against this Creature," and that belief held steady throughout the eighteenth century. In 1793, Thomas Jefferson maintained that "there is in man as well as brutes, an antipathy to the snake, which renders it a disgusting object wherever it is presented." In accordance with early modern beliefs that emotions played an active role in causing illness, fear itself could indeed be physically dangerous. For instance, Kalm relayed the firsthand account of a fellow Swede in Pennsylvania named Carl Låck who encountered a rattlesnake in his youth. Even though he ran away and thus avoided a venomous snakebite, Låck felt "so frightened, that he got sick and had to be helped home," where he lay ill in bed for fourteen days. One could say that this crippling fear proceeded from the human mind rather than any property of the snake itself. But such paralysis was not without foundation: report after report showed that rattlesnake bites hurt like hell, their venom causing an array of potentially lethal systemic, and often neurological, complications.[11]

One of the most haunting accounts of a rattlesnake bite and its impact on the human psyche appeared in the pages of the Royal Society's *Philosophical Transactions* in 1746. It took the form of a letter to Collinson from a Philadelphia naturalist named Joseph Breintnall. Breintnall opened his letter with an almost reproachful reminder to Collinson: "you injoin me a sad Task." In the somber report, Breintnall recalled being bitten by a rattlesnake and feeling "a Sort of Chilness when I heard the Sound; because I had a constant Thought, that if ever I was bit, my Life was at an End." Akin to other eighteenth-century observers who believed their skin grew snakelike after a bite, Breintnall wrote that his arm "seem'd almost void of Feeling; yet would it work, jump, writhe and twist like a Snake in the Skin, and change Colours, and be spotted." Snakes reminded humans of how woefully animal they were and how little they controlled the Chain of Being. But worse than any physical sensation, Breintnall said, "the most surprising and tormenting were my Dreams." Before the bite, Breintnall dreamed peacefully, "being ever in some pleasing Scenes of Heaven, Earth, or Air," but afterward—when he was lucky enough to sleep—he "dreamed of horrid Places, on Earth only; and very often rolling among old Logs," which "cast a sort of Damp upon my waking Thoughts, to find my sleeping Hours disturbed with the Operation of that horrid Poison."[12]

Though Breintnall was clearly alive and breathing when he wrote to Collinson, he concluded the letter by deeming it a narrative of "the fatal bite," and he drowned in an apparent suicide the following month. The rattlesnake's venom acted as a mind-altering agent, such that dreams became the stuff of science, published by the premier scientific journal in the English-speaking world. Snakes compelled multisensory engagement from their human observers, generating natural histories of fear rather than distanced, reasoned reports. Sounds, smells, disgust, pain, prejudice, even nightmares appeared in scientific accounts of rattlesnakes alongside more typical anatomical, visual, and behavioral descriptions. Human frames—and, by extension, human minds—were indelibly implicated in the study of serpents.[13]

Such ophidian fears made Euro-Americans uncomfortably aware that they were not immaterial observers with sovereign power over nature: they, too, were animal bodies engaged in acts of mere survival. The ability of snakes to seemingly alter human minds and human constitutions contributed to an overabundance of information and speculation about the animals alongside increasing doubt as to their nature. Benjamin Smith Barton, who had proclaimed the study of snakes to be uniquely prone to ignorance, attempted to write a magnum opus on American serpents and on rattlesnakes in particular. Yet he found himself paralyzed by an acute fear and loathing of snakes, a "weakness" that he felt was "the only prejudice which, I think, I have not strength to subdue." In fact, Barton's terror delayed his research agenda for nearly two decades. Once he finally worked more earnestly to surmount his aversions and even gingerly helped a captive rattlesnake shed its skin, he died before his ambitious rattlesnake book made it to print, despite advertisements announcing the work as forthcoming. The manuscript is now apparently lost, perhaps never finished at all. Rattlesnakes once again deferred human knowledge of their being—and in this case, they foreclosed it altogether.[14]

Did rattlesnakes, of their own accord, actually alter the minds and actions of the people who studied them? Or was it simply the severity of human reactions to these symbolically laden creatures that foiled attempts to document their natural history? Those questions gnawed at Oliver Goldsmith, the Irish-born naturalist and novelist who had also written at length about polyps in his encyclopedic compendium of natural history from the 1770s. Goldsmith noted the insoluble mixture of symbol, human perception, and actual danger in snakes. He observed a reciprocal relation between the serpent itself and human reactions toward it: the creatures were both deadly serious and metaphorically potent, which, he suggested, made them both colonial deterrents and elusive objects of science.

Repeating pronouns for emphasis, Goldsmith believed "not only their deformity, their venom, their ready malignity, but also our prejudices, and our very religion, have taught us to detest" snakes. Calling the serpent "the enemy of man," Goldsmith suggested that snakes' mastery of subterranean and other hidden worlds allowed them to contravene human projects to destroy them. "Formidable in itself," Goldsmith wrote in militaristic language, "it deters the invader from the pursuit; and from its figure capable of finding shelter in a little space, it is not easily discovered by those who would venture to try the encounter. Thus possessed at once of potent arms and inaccessible or secure retreats, it baffles all the arts of man though never so earnestly bent upon its destruction." Goldsmith took some comfort in the local extermination of many European serpent populations, which had made the life of the mind safer, allowing philosophers to "meditate in the fields without danger." But he identified snakes beyond Europe—whether the rattlers of the Americas, the vipers of Africa, the bushmasters of the Caribbean, or the cobras of India—as a prime obstacle to the spread of empire. "Nature," he wrote, "seems to have placed them as centinels to deter mankind from spreading too widely."[15]

Goldsmith ultimately doubted whether snakes, "undisturbed possessors of the forest," could be assimilated into natural history. The serpent appeared "too formidable to become an object of curiosity, for it excites much more violent sensations." In his opinion, those naturalists who did happen to procure the bodies of dead (and thus neutralized) rattlesnakes for their collections were missing the point. When Pehr Kalm visited Hans Sloane's then-abundant animal collection in 1748, he observed a stuffed rattlesnake skin and many other serpents preserved in fluid—perhaps including 1.11.1.a herself. Kalm was on an assignment from Linnaeus to count the number of scales and abdominal plates in Sloane's specimen of a hooded cobra sealed in a bottle, as Linnaeus's classification system entailed describing and quantifying an organism's external features. Kalm wrote of his maddening task with some spite: "So while the others went around and looked at everything, I had to spend my time trying to count [the scales] up, which was very difficult, since the snake was in a flask which was sealed at the top so that the alcohol should not become contaminated." He ended up having to hedge on the numbers. Goldsmith ridiculed such methods, insisting they missed the serpent for the scales. "Human curiosity," he wrote, "and even human interest, seem to plead for a very different method. . . . It is not the number of scales on a formidable animal's belly, nor their magnitude or variety, that any way excite our concern." Other naturalists found serpents similarly sticky to study neutrally, due both to fearful human reactions toward them as well as the animals' genuine penchant to escape the human gaze. Buffon, whose work heavily inspired Goldsmith, argued that large serpents such as rattlesnakes "are either too dangerous in reality, or are objects of too great terror, to admit of being observed with sufficient attention and perseverance." Human culture and the power of snakes themselves wove an unsolvable Gordian knot.[16]

Behold the rattlesnake, this reptilian marvel of sonorous music and underworld wiles—except, can you? A chasm of knowledge had opened up between people and serpents. Into it fell the natural history of rattlesnakes.

SUBTERRANEAN SLAUGHTER

While some naturalists were busy struggling to study this elusive creature and questioning whether they ought to do so in the first place, other Euro-American settlers turned to crushing the heads of serpents in biblical fashion. They worked to exterminate rattlesnakes at a time when humanity's ability to remove a link from the Chain of Being, by forcing a species into extinction, remained contested in scientific circles. The latter half of the eighteenth century proved a flash point in the environmental decline of rattlesnakes in New England, spurred by widespread determination to kill all rattlesnakes in sight, despite the animal's contemporaneous utility as a political symbol. Human hatred of rattlesnakes combined fatally with the underworld lives of the animals—indeed, what herpetologists, the scientists who study snakes, are increasingly characterizing as rattlesnake "culture"—to brutally deplete their populations. As a result, the snakes slipped further from historical memory.[17]

For many early Americans, killing a rattlesnake was the only acceptable outcome of any snake encounter. Even docile animals fell prey to this philosophy of destruction. John Bartram, who harbored a rare fondness for the creatures, recounted finding a rattlesnake coiled like "a great mushroom" that peacefully slipped away even when provoked. But Bartram's guide struck the rattler on the head with a stick and "said he never let any escape." For its last few moments of life, the placid snake, at last, began to rattle. Likewise, professional rattlesnake escapist William Byrd II noted the all-kill policy for even gentle animals, writing of how his expedition's surveyors on one day "kill'd two more Rattle Snakes, which I own was a little ungrateful, because two or three of the Men had strided over them without receiving any Hurt."[18]

Coordinated extermination took many more rattlesnake lives. Some New England towns organized community-wide hunting campaigns, and many offered bounties for rattlesnake bodies, as they would for other noxious beasts such as livestock-stealing wolves and crop-ravaging squirrels. Colonists who waged official extermination campaigns against rattlesnakes both framed and practiced their eradication as total war, going so far as to use the destruction of rattlesnakes as a military training ritual. In 1712, the Puritan minister Cotton Mather—equally famed for his promotion of smallpox inoculation and his involvement in the Salem Witch Trials—wrote, "The *Rattle-snakes* have their Winter-habitations on our Hills, in hideous Caves, and the Clefts of Inaccessible Rocks. In the Spring they come forth, & Ly a Sunning themselves, but still in pretty feeble circumstances." As the snakes were so exposed, Mather explained, colonial militias would then "take this time, to carry on a *War* with ye *Snakes,* and make ye killing of *them,* a part of their Discipline."[19]

Mather's barbed comments about these "hideous" and "Inaccessible" serpent hideaways betrayed an anxiety over their mastery of the subterranean, a realm far removed from the dominion of people. At the same time, however, the underworld social lives of rattlesnakes also provided a means by which humans could carry out their war. In autumn, timber rattlesnakes of the cold northeast return to their maternal dens, called hibernacula, and coil with one another for warmth and protection beneath the ground. Such subterranean habits made rattlesnakes difficult to study, from one point of view. But then, paradoxically, their sleepy mass exit from these communal refuges in spring temporarily made them highly visible and thus easier prey for widescale extermination.

Many eighteenth-century authors after Mather noted colonists' tendency to prey on this seasonal vulnerability by quashing rattlesnakes at devastating rates in spring and summer as the snakes warmed their bodies in the sun (figure 3.5). Kalm noted that rattlesnakes "congregate in their winter lair, crawling deep into the ground" and that in springtime "North American Europeans locate the lairs when these snakes come out en masse." The snakes were then still "sluggish" and easy to kill, as one Swede found when he dispatched at least sixty sunbathing snakes on a hillside covered in rattlers. Paul Dudley likewise reported that in Massachusetts, hunters would lie in wait, watching for the snakes to emerge and sun themselves, at which point they would slaughter them by the hundreds. Such numbers are likely not exaggerated: hundreds of snakes might occupy a single hibernaculum, and conservationists have documented that one notorious twentieth-century poacher killed or removed at least several thousand timber rattlers—sometimes more than one hundred

Figure 3.5. A painting by George Catlin containing the inscription: "The Rattle Snakes Den (fountain of poison) A scene of my boyhood in the Valley of Wyoming [Pennsylvania] Geo. Catlin." Catlin was born in 1796, which suggests that this incident occurred in the first or second decade of the nineteenth century. In the image, several hunters prepare to slaughter a community of rattlesnakes—likely containing many relatives—as the snakes emerge together from a den and bask socially. This seasonal community event for rattlesnakes often became a community event for humans as well; they would kill dozens and sometimes hundreds of serpents at once from a single hibernaculum. (George Catlin, *The Rattle Snakes Den [fountain of poison]*, 1852, oil on canvas, 18.125 × 27 in. [46 × 68.6 cm]; collection of the Gilcrease Museum, Tulsa, Oklahoma)

at a time from a given den—thus singlehandedly imperiling the populations in New York, Massachusetts, and Connecticut. In 1772, John Bartram connected the population decline of rattlesnakes to this practice of targeting dens, writing to Michael Collinson, Peter's son, that "ye reason of ye great decreas of ye rattle snakes is thair slow motion in runing away & thair gathering together from distant parts in ye fall to lodge together in ye winter where thay are found & multitudes of them killed." This transitional moment between the rattlesnakes' unknowable underworld existence and their terranean visibility gave humans a mechanism for exterminating creatures they otherwise longed not to encounter too closely.[20]

Allow for a moment of alterity. What might these scenes of destruction have looked like from a rattlesnake's perspective? In the dim light of the subterranean stronghold, any humans hovering near the entrance would appear to rattlesnakes as glowing globs of motile light. Specialized infrared-sensing pit organs on their faces (hence the term "pit viper") integrate heat data from warm bodies into a rattlesnake's field of vision. With this second sight, rattlesnakes might have also seen their kin gathered around if their bodies differed in temperature from the cool stone slabs, albeit at a lower intensity than any warm-blooded creature. Pheromones and vibrations, however, would be their primary tongue—in addition, that is, to their sensitive, flicking tongues. Humans may have had difficulty seeing inside the hibernaculum, let alone sensing these mechanisms. But their ears would tune to the chilling trill of a rattle if they ventured too close. The slithering, slow-moving, and, to the minds of some settlers, sacrilegious rattlesnakes who emerged from these dens would not be able to see with such infrared clarity the cool edge of a tool or gun barrel crashing down on their heads as they left their winter brumation.

Through targeting dens in this manner, settlers in effect preyed on the complex kinship and community structures of rattlesnakes, showing the importance of the animal's specific social and reproductive ecology to its population decline, and its illegibility to scientists past and present. Entire family lines of rattlers may have perished when snake families emerged from hibernation in spring to face colonists ready to slaughter them. Rattlesnakes were long viewed by modern-day biologists as solitary and unsociable. In revolutionary America, the rattlesnake was partly prized as a national symbol for being "solitary, and associat[ing] with her kind only when it is necessary for their preservation." Had that actually been true, timber rattlesnakes might have proven more difficult to drive to the brink of extinction in certain regions. Recent studies, however, suggest the rich social worlds of these creatures: rattlesnakes recognize and live with their relatives, they defend their young, and they warn each other of danger. In the words of the rattlesnake biologist Rulon W. Clark, "Snakes are often regarded as the least social of all vertebrate groups, but this assumption stems from the fact that they are secretive and difficult to observe in nature, rather than direct evidence." Rattlesnakes exhibit what biologists call "cryptic sociality," which has led their community dynamics to go unnoticed by scientists until very recently. The challenges of observing rattlesnakes in the wild, their modes of communication that rely on cues imperceptible to humans, and our own tendency to focus on the visual and vocal communication systems of mammals and birds all conspired to render the social lives of rattlesnakes largely invisible to the eyes of modern science.[21]

Though recent biological studies offer this imperfect window into the animal world, such research can still shed critical new light on historical interactions between humans and animals, especially when triangulated with other sources or used to read against the grain of early accounts. For instance, female social dynamics have had special import for the decline of the species. Female timber rattlesnakes like 1.11.1.a have recently been shown to recognize and preferentially cuddle their female kin even after years of separation. At the approach of warmer weather, pregnant rattlesnakes sunbathe together out in the open near their den entrance while others of their kind disperse into the cooler woods. Come August and September, these snakes will give birth to live young that they then care for. Given this prenatal practice, the vast number of settler reports that mention targeting groups of sunbathing snakes suggests many female animals in particular succumbed. Being gravid, and being gravid together, they were vulnerable. But removal of even a few female snakes, especially ones carrying young, spells trouble for a timber rattlesnake community. Some females take more than a decade to reach sexual maturity. They give birth only a handful of times throughout their long lives. Their litters are frequently small, at times bearing only three lacey neonates.[22]

Exacerbating these already steep reproductive pressures, European and American naturalists prized female rattlesnakes such as 1.11.1.a as sexual curiosities for public display and private study and dissection across the Atlantic world in the eighteenth century. The almost-human trait of live birthing in rattlesnakes—a feat uncommon among serpents, which typically lay clutches of eggs—fed a wider early modern hunger to observe reproductive oddities in exotic female beasts.

Thus, despite anxious Oliver Goldsmith's skepticism toward snake curiosities, many rattlers did appear in museum collections or public exhibits. It was a form of utility that countered their damnation as a discommodity. Given their fascinating anatomical features, rattlesnakes entered collections such as Hans Sloane's assemblage of 525 serpents either as whole animals or in bits as disjointed skins, fangs, skeletons, or lone rattles (figure 3.6). Some were dried, some stuffed, and some, like 1.11.1.a, stored in spirits, occasionally coiled around the edges of a bottle to fit, in the manner a potter would lay wet clay for a vase. William Byrd II had a penchant for using the uniquely American

Figure 3.6. Rattlesnake tail and rattle from the Drottningholm natural history collection of Queen Lovisa Ulrika of Sweden, mid–eighteenth century. (Photograph by Anders Silfvergrip; courtesy of the Swedish Museum of Natural History under a CC BY license)

rattlesnake, his surveying nemesis, to display his scientific prowess abroad. In 1697, along with a pouch-bearing opossum, he showcased a live rattlesnake—which he apparently had not fed for seven months—to London's Royal Society, and he later donated a rattlesnake's sloughed skin. At the same meeting, the apothecary James Petiver, who you will recall obtained much of his natural history collection by way of shark bellies and the slave trade, also unveiled a large and pregnant snake from Virginia. Petiver preserved her in rum, but not before extracting seventeen fetal snakes, one of which he also presented. Sloane would inherit Petiver's collection upon his death.[23]

Female rattlesnakes and their reproductive powers came to be standout attractions of private and public curiosity exhibits on both sides of the Atlantic. In 1738, one breathless London advertisement promised to exhibit in all its glory a "Female RATTLE-SNAKE alive and full of Vigour, the largest and most beautiful that has been in Europe within the Memory of Man; together with nine young ones, which came from her in her Passage to England; Curiosities never before known in this Part of the World, and which prove this Snake to be of the Viviparous Kind." Another ad suggested the titillation involved in such displays. Alongside a rattlesnake and her two young, the advertisement described the display of a young female chimpanzee brought live from Angola, along with hundreds of enslaved people, on the ship the *Speaker*. This "great Curiosity" had died the week before the display but was "so well preserved in Spirits as to give an increasing Satisfaction to those who have seen her when alive, as her Dress was then a great Obstruction to the handling and examining of her Neck, Shoulders, Arms, Breasts, &c. which upon the strictest Inspection of the greatest Anatomists, has confirm'd her to be the most extraordinary Creature ever brought into this Kingdom." In death, she could be fondled in peace by men of science bent on peering into the secrets of animal women; Hans Sloane was in fact one of her voyeurs in life, observing the ape and finding himself "extremely well-pleased." Both female rattlesnakes and chimpanzees fell under the male gaze as distinctly sexed and sexualized curiosities—though not only the male gaze. This advertisement also promised, as propriety would dictate, that a gentlewoman would show such female wonders "to the Ladies alone."[24]

Even if the rattlesnake's violent nature had convinced some naturalists to look away, when exotic females were involved, curiosity's eyes still peeked. The gendered assault on rattlesnakes, evident both in the extraction of pregnant females as scientific specimens and the massacre of clusters of sunbathing mothers, hastened the demise of the species by removing pivotal matriarchs from those snake societies.

ALLIES IN OBFUSCATION

European settlers did not only have to deal with snakes in matters and slaughters subterranean. For they were not the only people with a stake in rattlesnake societies, nor the only ones stuck in a careful-toed waltz between knowledge and ignorance when it came to the animals. Many Native peoples in early America were sensitive to serpentine family lines. Some counted rattlesnakes as kin or engaged with them in what were effectively political contracts. They had long recognized and documented the complex collective networks and bonds among rattlesnakes that herpetologists are only now, centuries later, beginning to see.[25]

Native peoples were also familiar with the concealed worlds of snakes, which played an important role in their relations with the animals, as evident, for example, in rattlesnakes' association with the Lower World or Under World in southeastern Native cosmologies. Various Indigenous peoples approached rattlesnakes through a lens of interspecies obligation and, in certain cases, calculated avoidance—dynamics that had their own implications for the survival of the timber rattlesnake. Working in tandem with the snakes' elusive nature, Native peoples provided another check on the Enlightenment quest to catalogue the nature of new worlds, even as they also contributed valuable insights and specimens to the natural historical project.[26]

One of the most detailed Indigenous accounts of rattlesnake kin networks from early America appears in the work of Cephas Washburn. A Presbyterian missionary, Washburn recorded Cherokee oral histories in the early nineteenth century. One interviewee, a Western Cherokee man named Blanket, shared a rattlesnake story from his youth in the eighteenth century. (He was around eighty years old at the time Washburn interviewed him.) After Blanket and Washburn encountered a snake one day, Blanket physically turned away and refused to harm it. Washburn, by contrast, promptly ordered his interpreter to slay the animal. Blanket said his philosophy for dealing with serpents was "I never kill snakes, and so the snakes never kill me," intimating reciprocal obligations between humans and rattlesnakes through mutual recognition and avoidance. Blanket's rich elaboration of the encounter behind this philosophy, presented by Washburn as an extended direct quotation, reveals an alternative eighteenth-century approach to the rattlesnake's world: one oriented vertically as opposed to the horizontal vantage of settler expansion and surveying, and one that foregrounded animal kinship and subterranean sociality.[27]

As a young man in the eighteenth century, Blanket traveled with a companion who killed a rattlesnake. The following day, a rattlesnake emissary met Blanket in the woods and invited him through an opening in some rock and then down into the "subterranean abode of the serpent race." Blanket saw the body of the snake that his friend had killed stretched before a fire in a great underground hall, while the snake's "father and mother and all the near kindred of the dead snake were making great lamentation over it." Each snake said its last farewell, and they concluded the funeral rites by interring the snake with other fallen comrades. Next, Blanket witnessed a council of rattlesnakes debate how to secure justice for the slain animal. They agreed that Blanket's human companion must be killed. The importance of rattlesnake family ties became even clearer when they selected the murdered snake's brother as his avenger—a choice that mirrored the kin-based revenge practices of various Native nations, here decided collectively with "a hiss of approbation, accompanied by a concert of all the caudal rattles of the assembled multitude." Blanket confessed to his rattlesnake guide, "I could but acknowledge that it was even-handed justice, and in accordance with our customs, in similar circumstances." Within days, Blanket's friend died from a rattlesnake bite to the heel.[28]

Blanket sensed his own status as a guest in that space. Although a single account could never capture the tremendous diversity of Native viewpoints on rattlesnakes, his encounter illustrates obligations between Cherokee individuals and rattlesnakes not to kill one another—obligations of mutual protection that held true for several American Indian nations. Blanket's account, paired with the divergent reactions of Blanket and Washburn to the snake who crossed their path, also begins to hint at how Native American understandings of rattlesnake underworlds hindered European efforts to extirpate the creatures.

Many Native peoples fiercely guarded the whereabouts of rattlesnake societies as secret knowledge. James Adair, a deerskin trader who worked with various southeastern Indigenous communities, said of Native attitudes toward one reported domain of giant rattlesnakes, "They reckon it so dangerous to disturb those creatures, that no temptation can induce them to betray their secret recess to the prophane. They call them and all of the rattle-snake kind, kings, or chieftains of the snakes." Native people also shielded individual snakes from harm. For instance, Adair noted that he once saw a Chickasaw man "chew some snake-root, blow it on his hands, and then take up a rattle snake without damage," at which point the man hid the rattlesnake in a hollow tree to prevent Adair from killing it. Kalm mirrored Blanket's language when commenting on the population collapse caused by "the merciless extermination of the rattlers by the Europeans" and observing that a Native individual "who comes upon a rattler does not disturb it, but says in passing, 'You go your way and I'll go mine.'" And the Moravian missionary John Gottlieb Ernestus Heckewelder similarly wrote to Benjamin Smith Barton about his experiences with the Lenape in present-day Ohio: "In the Year 1762 . . . I was several times prevented by Indians from killing them, being told that the Rattle Snake was their Grandfather, & that he durst not be hurt."[29]

All of these vignettes come to us from settler sources, which often misunderstood or misrepresented Indigenous practice. Nevertheless, these recurring moments of mixed obfuscation and protection across a wide geographic expanse collectively suggest that numerous Native nations strategically safeguarded rattlesnakes from annihilation. And they very often did so by building on the animals' own penchant for concealing themselves. Native people certainly did kill rattlesnakes for purposes of food, dress, technology, and self-preservation. In fact, some Euro-Americans hoped to recruit Native individuals in their eradication attempts. But the efforts of many to protect rattlesnakes provided a significant check to settler extermination efforts.

The sense of danger infusing some of these accounts, however, suggests that Native people did not necessarily obscure the world of rattlesnakes out of proto-conservationist sentiment. Rattlesnake avoidance could preserve Native communities by forestalling the retribution of vengeful snakes—even if this avoidance ultimately did result in the salvation of rattlesnakes, either through keeping settlers ignorant of den sites or protecting or hiding individual animals. Additional sources speak to further complexities of these snake protocols among Native peoples. Multiple accounts, for instance, show that some Native Americans literally defanged the snakes to reduce lethal bites while still letting the snakes live, thus technically fulfilling those mutual obligations to safeguard one another. Adair told of one tragic encounter with an "old Indian trader, inebriated and naked, except his Indian breeches and maccaseenes," who had smeared his hands with medicinal roots in preparation for wrenching the fangs from the mouth of a large rattlesnake he was holding. Upon completing his objective, he laid the snake "down tenderly at a distance," hoping he had achieved peaceful coexistence. But Adair promptly killed the serpent "to put it out of misery." This, however, was to the man's "great dislike, as he was afraid it would occasion misfortunes" to both himself and Adair. Moreover, as Maurice Crandall, a historian and citizen of the Yavapai-Apache Nation, told me of present-day southwestern Native communities, avoidance of rattlesnakes in that part of the continent does not normally equate with protection at all, given the taboos associated with the creatures. Rattlesnakes are viewed negatively—and killed, if encountered—partly since they are subterranean and thus adjacent to buried ancestors.[30]

In some eighteenth-century cases, Native Americans might even outsource rattlesnake killings. One Seminole community asked William Bartram to slaughter a large rattlesnake that had entered their camp. Their proposal seemed a win-win: they would avoid rattlesnake revenge for themselves, and, they reasoned, it was Bartram's "pleasure to collect all their animals and other natural productions of their land." In this instance, Native people led naturalists *to* rattlesnakes, their snake avoidance in fact aiding natural history instead of denying it. Living with and understanding rattlesnakes, these literally and cosmically potent beings, was never simple or straightforward for any human society.[31]

And in time, William Bartram swore off killing rattlesnakes for good.[32]

EXTINCTION AND DISENCHANTMENT

The death of animals—along with its most extreme form, extinction of a species—posed direct relevance to natural history's mission to understand creatures like rattlesnakes. Take them away, abuse them, reduce them, deplete them, and how can one claim to know them? Naturalists debated the reality of extinction throughout the eighteenth century. Within a worldview bent on discovering a purpose-driven nature, extinction's finality would imply that God's perfectly gradated Chain of Being could be ruptured by environmental change, or even by human will. So said Jefferson, a true doubting Thomas: "Such is the economy of nature, that no instance can be produced, of her having permitted any one race of her animals to become extinct; of her having formed any link in her great work so weak as to be broken." The French anatomist Georges Cuvier established extinction as an undeniable feature of nature while working with the teeth of extant elephants and extinct mastodons and mammoths as the eighteenth century bled into the nineteenth. But throughout the longer early modern period, the decline of more recent species—famously, the dodo of Mauritius—and the noticeable impact of humans on colonial environments over short timescales made humans' ability to indelibly transform the natural order a serious, if controversial, possibility.[33]

In the early 1770s, John Bartram (William's father) and Michael Collinson (Peter's son) exchanged several letters about the prospect of rattlesnake extinction. Collinson desired their annihilation but doubted its feasibility. He wrote to the elder Bartram: "The Extirpation of that dreadful Animal the Rattle Snake will be never accomplished notwithstanding the perpetual war against the Race—the Continent of America is so vast, the Retreats so many—and so secluded from human Approach, that if it ever should take place it must, I think be many Centuries first." His comments were tragically prescient, as the snakes skirt extinction in certain regions today, right on schedule. But Collinson's death wish for rattlesnakes also reveals an implicit lay belief in the physical possibility of extinction, despite the opinions of naturalists like Jefferson. Collinson framed the question of extinction as a matter of whether one could know and surveil rattlesnakes well enough to exterminate them—in this instance, perhaps not, or at least not soon, due to the animals' mastery of secluded spaces and the expansiveness of the North American continent. Those hideaways that made rattlesnakes vulnerable in springtime also pocketed them away and made their full extinction impractical, as Goldsmith had despaired. Rattlesnake extinction, for Collinson, was a failed colonial experiment, but one that might succeed under the right conditions.[34]

Bartram had quite a different take from his correspondent. He outlined the grave consequences of extinction for understanding the powers of animals, insisting that extinction was what produced myth, driving a warped conception of former life. "Ye notion of basilisks, Dragons & unicorns," Bartram told Collinson, "is disbelieved by many never to have existed on our globe but I shall not deny thair being in considering how many remains of many creatures is frequently found in our days which have existed tho not living within ye reach of history." Bartram wrote these words in 1772, at a moment when people were unearthing remnants of extinct beings, from giant mastodon bones to branching Irish elk antlers. He reasoned that so-called mythical beasts such as dragons and unicorns might actually have existed once, just like these other now fossilized creatures. They simply vanished from the annals of history. Extinction led to disbelief and disenchantment.[35]

What then, Bartram wondered, would the depletion of the animals *he* studied mean for future historians of his own time?

In the case of the imperiled rattlesnake, Bartram feared the creation of an American fable. "Our Bufeloes beavours & rattle snakes," Bartram wrote in alarm, "is so like all to be distroyed that if thay are so continualy destroyed for A Centry to come as thay have been for ye last 50 year there will be but few left & ye story of A rattle snake & its fascinating power will be in A few hundred years as little credited as ye cokstrice [cockatrice, a legendary serpent] killing at A distance by its eyes." Bartram had witnessed with remorse the decline of rattlesnakes over the course of his more than seven decades on earth. His allusion to their "fascinating power" referenced a contemporaneous debate as to whether snakes could "fascinate," "charm," or "enchant" prey—or even humans—with merely a hypnotic glance or a shake of their rattle, causing a victim to freeze in its tracks. Bartram mourned not only the physical loss of animals to extinction but also extinction's threat to erase history. Time, he believed, would attenuate the power of animals and leave stories of their behavior in the realm of mere fairy tales.[36]

In the eighteenth century, anyone who had an opinion on rattlesnakes—

Barton, Bartram, Byrd, Catesby, Collinson, Dudley, Kalm, Lawson, Sloane, and countless others—had an opinion on rattlesnake enchantment. Although the notion applied to snakes more broadly and found precedent in earlier European, African, and Asian thought, it took on renewed life and reached its apotheosis in the loaded and lethal American rattler. Some located the seat of this terrifying power in the undulating colors of the serpent's body; some in "the dazzling brightness of their eyes"; some in the sound of the rattle, "the most rapturous strains of music, wild, lively, complicated"; some in what they believed was the rattlesnake's willfully emitted fetid odor. And enchantment surfaced in sundry New World contexts, showing a deep American past. For example, the sixteenth-century Mesoamerican Florentine Codex, a collaboration of the Franciscan friar Bernardino de Sahagún and Nahua scholars, intriguingly notes that rattlesnakes were known to shoot "something like a rainbow" from their mouths to make a chattering squirrel swoon and fall unconscious from its perch atop a cactus—yet another variation on the theme of action at a distance. Still more theories jockeyed to explain the phenomenon in the eighteenth century, with some blaming fear itself.[37]

Hans Sloane himself had risen to prominence in this discussion during the 1730s by providing one of the first supposedly "scientific" explanations of the phenomenon. After observing firsthand the effects of a rattlesnake's fast-acting venom when it sank its teeth into an ill-fated dog, Sloane suggested that enchanted prey had simply already been bitten and paralyzed, which would explain their inability to flee from the snake. (Never mind the many humans who suffered something like enchantment but made no contact with the serpent; Byrd, for instance, claimed that he "Ogled a Rattle Snake so Long" that he grew sick to his stomach, proving to his correspondent, Peter Collinson, their "power over men.") Sloane's hypothesis ignited a sequence of theories that would ultimately neuter the snake of such powers, especially through the writings of Benjamin Smith Barton.[38]

Centuries after Bartram voiced his worst fears about what death might enact for rattlesnakes, historians, biologists, and other scholars and writers today tend to deny that rattlesnakes ever had any real ability to fascinate. They deem it a myth and occasionally go so far as to fault the many eighteenth-century naturalists who believed it as overly credulous. Some dismiss enchantment as "outdated" or "not true" and move on; some concede that humans from centuries ago might well have been fascinated, but "not, needless to say, by any power of a snake." It is always taken as a given that rattlesnakes had no efficacy of their own.[39]

Bartram was right.

Rattlesnakes went from being formidable threats with the power to suspend European understandings of American nature and landscapes in the seventeenth and eighteenth centuries, to the object of extermination efforts that furthered the loss of rattlesnake knowledge by nearly eradicating the animals themselves in many regions, to creatures increasingly associated with myth as their numbers dwindled by the end of the eighteenth century. As cryptic animals with a secretive ecology that is still barely understood by scientists, as legless masters of the underworld whom Goldsmith branded "centinels to deter mankind," as rattling animals with something to say back to symbolizing settlers, as creatures with Native protectors, and as beings that can strike fear into people's hearts, rattlesnakes brewed a perfect storm of ignorance for colonial science. It is ever easy to lose our moorings when speaking of rattlesnakes. They were unbearably freighted with meaning. Yet I.II.I.a existed as an individual in this period. She remains, having escaped the bonfires that consumed the bulk of Sloane's other animals. Might she be a buoy for an otherwise submerged tale of loss? Might the subterranean—so to speak—speak?[40]

I've seen countless rattlesnakes jump out at me like specters from the pages of eighteenth-century books, journals, and manuscripts. The vast majority are uncolored, like I.II.I.a's pale scales. Each time I ask: could it be her? Each time there are no definitive answers. Sloane left behind a manuscript catalog of his serpent specimens. The entries alternate between detailed narratives and bare descriptions of individual specimens—sometimes acknowledging a specific donor or a brief story about the snake's life, but more often curtly noting something to the tune of "skin of a rattle Snake." The story of how I.II.I.a entered Sloane's collection is probably unknowable. Labels on her bottle that might have offered a number corresponding to Sloane's inventory are too mutilated and faded to interpret. Yet the archival record does leave a few suggestive clues as to possible identities.[41]

She could be that rattlesnake brought to England by William Byrd II, whose boundary survey was continually delayed by the very animal he converted into a natural curiosity. One entry in Sloane's personal catalog lists "a rattle snake with the mouth open brought alive into England by Mr. Byrd," suggesting one potential identity for I.II.I.a, whose mouth remains agape. Another clue in her preserved form sparks a more speculative suggestion. She could be one of the snakes who disenchanted the world, providing Sloane with the first purportedly empirical explanation of enchantment and setting naturalists on a path to question the rattlesnake's powers. Some bits of a gravel-like substance line the bottom of her present bottle. One rattlesnake who helped inspire his venom theory as an alternative to enchantment prompted Sloane to write: "Mr. *Read,* an eminent Merchant in the City of *London,* had a *Rattle-Snake* sent him alive in a Box with some Gravel from *Virginia,* which he did me the Favour to give me." Just as likely, however, she may be one of the many anonymous and quotidian rattlesnakes on Sloane's list, gathered by an unknown, enslaved, or otherwise unnamed collector. Though inconclusive, the exercise of weighing the possible identities of I.II.I.a by reading through Sloane's catalog and other sources forces us to see this snake as an individual, not just as the symbol humans asked her to be. By virtue of being a rattlesnake in Sloane's collection, she was, by definition, entangled in Enlightenment-era debates about the mysterious nature of these animals. We would do well to keep her in view even as her story remains inscrutable.[42]

Today, in regions that have seen a precipitous drop in the number of live timber rattlesnakes underfoot, a person will more likely encounter rattlesnake imagery from revolutionary America—on bumper stickers, "Join, or Die" tattoos, and Gadsden flags billowing at televised uprisings—than a moving and shaking rattlesnake in the woods. The descendants of creatures such as I.II.I.a persist even in regions of steep population decline due to a campaign of calculated ignorance. Conservationists now treat timber rattlesnake hibernacula as unutterable secrets. They sedulously conceal the location of snake communities because people continue to hunt down and sometimes exterminate societies of rattlesnakes simply for their crime of being rattlesnakes. The herpetologist William Brown devotes an entire section to maintaining secrecy in a conservation guide for the species, writing, "It is essential to never reveal the location of hibernating dens to anyone, except for valid reasons (e.g., research or protection)." He claims, "More than all other factors, it is this lapse of judgment on the part of those with this sensitive knowledge which has, over the years, led to massive exploitation." Many state-level efforts to conserve timber rattlesnakes promote the walling off of human and snake worlds as much as possible, a tactic that recalls the mutual avoidance strategy of Blanket. This modern-day practice is an unacknowledged legacy of Native American concealments of rattlesnakes from settlers in the eighteenth century. Ignorance and secrecy have thus served as something good, something noble, in the minds of historical and contemporary actors, counterintuitively protecting environments. Saving rattlesnakes continues to rely on policing access to knowledge.[43]

The serpents themselves have a part to play too. Their secretive nature, which helped rattlesnakes survive the slaughters of the eighteenth century, could be adapting to become even more elusive. A few early studies suggest that certain species may be losing their rattling ability or that those still able to rattle might be rattling less frequently as a coping mechanism to avoid human detection and human violence. That the rattler's warning, interpreted by early Americans as a miraculous gift from a wise and just God, could abate in the face of human onslaught seems poised to make the rattlesnake itself one of John Bartram's mythical beasts: an animal oracle who tried to warn humans of coming harm. We might listen while we still can.[44]

CHAPTER FOUR

Secrets

THE ENIGMATIC I.II.I.A HAS BEEN, for me, akin to the rattlesnake emissary who led Blanket into the serpent subterranean—a guide for exploring a dark and unfamiliar place, who, in an inversion of Plato's allegory, showed me the cave itself, alerting my senses to a realm underfoot that I had barely registered before and that still impressed on my mind only as faint shadows.

Imagine my surprise when I learned that I had received official authorization to meet the living heirs of the historical serpents I studied, to observe the rims and edges of this underworld. Timber rattlesnakes, the blood of I.II.I.a, are the most critically endangered animals in New Hampshire, my place of residence at the time. Only one known population remains in the Granite State. With that in mind, the New Hampshire Fish and Game Department, self-described "guardian of the state's fish, wildlife, and marine resources," keeps the whereabouts of these rare creatures a virtual state secret even to most of their employees. They scrub the den's location from maps in official documentation and police physical access to it—although, I'm assured, the guardians won't deceive if you do discover the site. Certainly, there are others in these woods who know of it.

Wildlife biologist Brendan Clifford is one of the state's appointed silent caretakers, sworn to keep the snakes hidden and manage their well-being. When I first spoke with him, over the phone, I learned he had been monitoring this population for nearly fifteen years by tracking their seasonal movements, recording the ebb and flow of their births and deaths, observing the ravages of a fungal disease plaguing the population, and rehabilitating injured and sick snakes. His Bartramian tenderness toward these animals sat in stark contrast to many eighteenth-century settlers and naturalists who either avoided rattlesnakes or set out to annihilate them. Brendan surmised that for some of the serpents in this isolated community, he was the sole human being they had ever seen. One contemporary nature writer, I learned, had already been blacklisted in herpetology circles for not taking sufficient care on the printed page to keep the location of this and other rattler dens unknowable and untraceable. Despite the profession's unease with authors, after our phone conversation, Brendan's supervisor, for whatever reason, granted me permission to follow Brendan to the hibernaculum.[1]

Since Europeans first invaded this landscape, the ancestral and still-unceded homeland of Pennacook, Abenaki, and Wabanaki peoples, rattlesnakes have had figurative targets on their cursed backs. Current secrecy efforts help guard the animals from an extermination ethos shared by many early settlers: a belief in a God-given duty to kill rattlesnakes as a satanic stain on an Edenic landscape. Brendan informed me of a snake slayer named Rudy Komarek, deceased as of 2008, who remains notorious among the conservation community as a "fanatical enthusiast" and federally convicted wildlife trafficker. He singlehandedly removed thousands of timber rattlesnakes from dens in the northeastern United States, often under the guise of protecting public safety. Komarek killed snakes, traded their limp bodies for bounty payments, sold some alive into the commercial trade, left some for dead in sealed bags as he evaded wildlife officers in hot pursuit, denied water and care to others for so long that their bodies aborted their babies, stillborn. He reportedly sold maps of den sites to like-minded collectors in a direct reversal of the conservationist practice of geographic occlusion. "In the snake community," Brendan said, "he was the number one guy to look out for." Herpetologists William Brown, Len Jones, and Randy Stechert had harsher words for Komarek in a co-authored journal article, branding him a "pathological snake-hunter" on a par with the settler-exterminators of early New England, a "nefarious hominid" who "seems even to relish the idea of his power to exact a devastating toll on this species." Early America, I learned, is never far from the minds of timber rattlesnake biologists when dealing with the likes of Komarek. They look back to the eighteenth century and its "purposeful snake-hunting raids" that exterminated historical dens as the first sins in a chain of events that culminated in the mess they must work to clean up today, assured that they are themselves part of a different lineage of rattlesnake investigators.[2]

But Brendan wasn't keeping the den secret from only the Rudy Komareks of the world. He told me how the merely curious, people not unlike myself, seeking to observe these rattlesnakes either for their extreme rarity or for the unique black coloring of this particular population, could cause the community distress and unintentionally expose the site's whereabouts to others. On its surface, my curiosity, or Brendan's, or that of an anonymous snake peeper seemed distinct from the eighteenth-century breed. We don't tend to think of ours as so volatile, hovering between observer and observed, liable to lapse unpredictably into one or the other without warning—though we will soon see that it can, we humans being still not so far from objecthood as we might like to think. Certainly, however, the danger curiosity might pose to snakes has resonated across time. I thought of the Enlightenment-era investigators who captured these animals not only out of fear or a desire to enact revenge on the species but as a curious pursuit—to study and dissect their bodies and convert them into specimens and images or to parade not-yet-dead snakes and their live-birthed young in public displays. The state's recent efforts worked to sever human and snake worlds, curiosity be damned.

To see these secretive serpents, we were beholden to their own schedule and rhythms. We counted down the days until warmer weather would coax the vipers en masse from their winter hideaway, many fathoms deep in dark fissures of rock. We waited. And we waited. Once the snakes emerged from their hibernaculum, they would bask in the sun to raise their body temperature and then disperse miles into the woods for their short summer window of terranean activity. Postpartum mothers who had given birth to litters of live young the previous year would continue the slow labor of rebuilding their fat stores after their resource-intensive pregnancies. And those females in the population who had mated in a dance of intertwined tails last autumn, who had been holding and quietly tending sperm in their bodies all winter, would now begin their formal gestation at the very moment we sought them.

Only, as luck would have it, this particular year was the infamous 2020. Our snake community's spring exit happened to coincide with the early days of the Covid-19 pandemic, wrought from some unknown form of human-animal intimacies. I worried our expedition would be delayed indefinitely on account of the strict lockdown. But despite the virus raging around us, care of these endangered animals was deemed essential state business, given their precarious hold on existence.

When I told my friend Charlotte about the upcoming trip, she asked an offhanded question that I couldn't shake from my mind: "What's more dangerous: A stranger in a mask or a mama rattlesnake?"

The mama rattlesnake, I thought to myself without hesitating—but perhaps not for the reason you might answer the same. I feared a masked stranger, but I had faint premonitions of what a mother was capable of. For I, too, was pregnant.

I hadn't left the house in months, apart from scaled-back prenatal visits to the hospital. Faced with the prospect of venturing into an apocalyptic world, to root around for a potentially lethal animal across steep terrain, with a creature in my body I was determined to care for, and, as I saw from the ever updating weather report, all during a massive heat wave, I began to feel some apprehension. I debated whether to reveal my pregnancy to Brendan in advance for fear I would be banned from seeing the den, my state of being too much of a risk for the state of New Hampshire; and then, when I decided to tell him and he was happy to proceed, I felt trepidation that this trip was actually going to happen, only days before I would enter the turbulent third trimester, as my lungs started to lose precious capacity and my abdomen and sense of balance grew yet more unwieldy. Would I react like William Bartram, I wondered, who, when faced with one rattlesnake, felt "so shocked with surprise and horror as to be in a manner rivetted to the spot, for a short time not having strength to go away"?[3]

Before I left in search of snakes, my father-in-law, Guillermo, offered a bit of family lore—something we call a Robles Story. A woman who worked at the ranch and general store of Guillermo's grandfather in Veracruz, nestled along Mexico's Gulf Coast, also encountered a rattlesnake as an expecting mother. Early one morning, while she milked a cow in the semi-dawn, a serpent hiding behind the livestock struck out from the shadows and envenomed her leg. The woman cried out in pain, only to find, to her surprise, that the rattlesnake died instantly and collapsed in a limp spiral at her feet. A physician was called to the ranch to treat the woman. But when he arrived, he told her not to fear: he insisted that the forces of vitality flowing through her body from the pregnancy had simultaneously killed the serpent and protected her unborn child. He assured her: "There is nothing more poisonous than a pregnant woman." After a grueling labor several months later, she gave birth to a baby, who was unscathed. And out with the child came a tidy sealed sack of rattlesnake venom.[4]

The morning of our journey, I found William already on the couch downstairs. He had woken hours before me due to a serpentine knot of nerves in his stomach, which he hoped either to ease or exacerbate by watching videos of timber rattlers, looking for some sign that these beings would not harm his wife or child. Brendan had given William permission to join the excursion since social distancing requirements made Brendan unable to help a substantially pregnant woman scale a cliffside. As William and I started our drive shortly thereafter, a large and surly snapping turtle lumbered across the road and temporarily blocked our path, perturbed at the humans who dared help it to safety—as good a reptile omen as any. The thermometer in our car soon read 95 degrees, highly unusual for a New Hampshire morning in late May. It was yet another reminder of our tinderbox of a world, which seemed ready to ignite on all sides from pandemics and melting permafrost and police brutality, as protests over George Floyd's murder would erupt in just a few days, his body not yet laid to rest beside his mother, inside the earth. The trees on distant hillsides shimmered sickly from the heat's haze. I had stocked our hiking bags with as much water and Gatorade as would fit, knowing we wouldn't be able to stop at so much as a gas station, thanks to the virus's insidious hold on potentially any indoor space. Nature would be our rest, and our restroom.

We drove to a location I cannot speak of, passed trees whose nomenclature I should not name, leaped—as much as the gestating can leap—across rocks whose form and geologic history must go unsaid. The rocks, I will note, were towering and magnificent, ancient wardens sheltering legless spirits in their stony wombs.

But there are other things I can't tell you.

We initially greeted Brendan from afar in a parking lot. He asked us to follow his vehicle, which made strategically confusing twists and turns, to the undisclosed-to-us location. As we began our hike, Brendan mentioned feeling anxiety about the upcoming workweek. He'd be stationed in a park on New Hampshire's coastline to protect state-endangered piping plovers, which had nested along the beach just as tourists would try to descend for the summer. He worried for the plovers and would plant signs to shoo away beachgoers; he also worried about exposure to the virus, which he feared bringing home to his medically at-risk daughter. We all agreed to stay the hell away from each other in the friendliest way possible—though we almost broke protocol once when Brendan excitedly fell to the ground to catch a spring peeper and extended his cupped hands to show us the tiny frog, until we realized what was happening and all jumped back, amphibiously.

We likewise planned to socially distance from any rattlesnakes who might cross our path. Increasingly alarmist signs on our ascent warned "Stay on Trail or Stay Home"—the forest's own version of "Don't Tread on Me." But with the blessing of the state, we diverged from the sanctioned path, entered the brush, and let the bushwhacking begin.

I actually did own something like "wide sailor's pants," Pehr Kalm's outfit of choice for navigating rattlesnake territory in the eighteenth century. (Kalm also mentioned that people in labor draped rattlesnake skins over their bodies "to promote an easy delivery," though I wasn't yet sure such a thing existed.) Opting for modern protections that day, I instead wore maternity jeans covered by snake gaiters that Brendan had supplied to provide bite protection from ankle to knee. I also donned a white cotton shirt, good for keeping cool and for making black-legged ticks, spreaders of Lyme disease, simpler to spot. Given the distinctive onyx hue of the snakes in this population, known as black morph or black phase as opposed to the yellow-brown coloring of most timber rattlers, Brendan told us to be on the lookout for anything that resembled a clump of moose poop (not that I was terribly familiar with the sight). They especially liked to hide along the edges of fallen trees, he warned. With each step, I surveyed the land past the hill of my pregnant abdomen to ensure I didn't step on a coiled snake.[5]

By this point, readers from the American South or West, or from many regions of Latin America, or simply those with a cherished phobia of snakes may see me as woefully naive. Timber rattlesnakes—especially the critically imperiled populations throughout the Northeast—are only one small sampling of the diversity of rattlesnakes in the Americas, totaling around thirty different species. Some rattlesnakes, like diamondbacks, come into frequent and bold contact with humans. But Brendan informed me that this sheltered and fragmented population in New Hampshire bears so little ill will toward the foreign form we call the human that they don't typically even rattle when he approaches. For all I know, they could be entirely behaviorally distinct from their ancestors who snaked through these forests in the eighteenth century.

I daydreamed that we would find the rattlesnakes emerging together from the rocks, their remarkable reptilian social structure on view. To survive, they must join, or die. But once we reached their stone stronghold, we realized the snakes had beaten us to the punch. Already dispersed into the woods, they likely would not return to the den until later in the fall or, in the case of pregnant females, a bit sooner to begin a gestational sun vigil together. These snakes were not the writhing spectacle of vulnerability I expected to find, which had so exposed them to settler predation in early America. In fact, tracking devices that had been affixed to four females provided the only clues to the existence of any snakes at all in these quiet woods. Brendan's team had placed radio transmitters either directly onto their rattles with glue or, more invasively, through surgical insertion at a veterinarian's office. Each snake had her own FM wavelength, and to track her down Brendan would hoist a giant metal antenna above his head—similar to the ones you'd once expect to find on the roofs of suburban homes—and follow a steady beeping that intensified as we ranged closer. In addition to listening for their quivering rattles as so many eighteenth-century wanderers had done, we readied our ears for very different soundwaves.[6]

As we moved deeper into the forest, I felt not fear but a surprising alertness and calm. I silently repeated a mantra that would carry me through this miles-long search over and through bramble, steep cliffs, slippery leaves, mosquito clouds, and menacing logs to meet the one I sought, a mantra I would repeat again, just months later, throughout my own labor: "There is nothing more poisonous than a pregnant woman."

The rattlesnakes and I—we had all been waiting. Life had been paused, dormant, latent, on hold. Timber rattlesnakes, near the northern limit of their range in New Hampshire, spend the majority of every year underground, deep below the frost line. Brendan hoped to acquire a borescope, a camera on a wire dozens of feet long that could explore the network of stony channels hosting this community. But without the human technological snake, this pit of pit vipers stayed hidden in darkness, their underworld lives perhaps better known to Blanket than to modern-day herpetologists or to me. Many female rattlesnakes spend these dim subterranean months readying themselves for a spring and summer gestation. In summer, pregnant females join together as if to form one mass, enveloped in each other's coils and folds while exposed on sun-soaked slabs of rock near their maternal den, one more instance of the firm sense of home ingrained in the species. In late August or early September, these females will give birth to reptilian replicas in miniature as the surrounding treetops burst into fireworks of redback salamander and burnt orange—the very same season when I would be due to deliver a daughter.

So while this community of snakes had been waiting, I was waiting out a biological and cultural winter of my own. In early March of 2020, I retreated into my own den, into my own self, for quarantine, all while tuning my senses to the small life growing inside me. And then March became April became May. Time twined, knotted, rolled over itself and concealed its true shape. I felt my child's first tadpole twitches in this state of suspended animation; early modern women called a fetus's initial perceptible taps "quickening." Pregnancy wasn't the worst time to slow down and simply be, for those like myself with the privilege to afford stasis. I was curled up, becoming, snake-like in my incubation, expecting, expectant, waiting for things not yet seen.

And yet, the early season of this pandemic was one of the worst times in recent historical memory to be with child. To have little choice about bringing new life into a major historical event, as I was already pregnant when lockdown began and soon realized my daughter would enter a world quite different from the one she was planned in; to know in your mammal gut (and from basic logic) that gestation put you at a higher risk of severe illness and death, only to be dismissed by doctors and federal agencies until "the needed data" vindicated you long after it mattered; to lose any human touch that wasn't your lover, save for a nurse's chilled hand; to wonder if a vaccine would ever materialize; to flee at the sight of people, given the very real consequence, should you or your partner be infected, of birthing a baby alone—a baby that would then be whisked away from you for days or weeks, under hospital protocol; to miss out on any sense of normalcy before such a metamorphosis in one's life. All this, to emerge and suddenly find that the village you were promised is empty.

Scientists do not typically describe rattlesnakes, or other reptiles for that matter, as pregnant. Instead, they are termed gravid, from the Latin *gravis,* meaning "burdened" or "heavy." The word initially struck me as more clinical, less human—an attempted bulwark against anthropomorphism. It was not until I became pregnant that I would realize just how clinical and yet human it could be. When reviewing my medical records, I noticed the label G1P0 on every document: G to denote gravidity, the number of times a person has entered this state of burden and heaviness, and P for parity, defined in the United States as the number of previous pregnancies that made it past twenty weeks of gestation. Various pregnancy-induced medical conditions, like the brutal onslaught of chronic vomiting known as hyperemesis gravidarum, likewise imbue human pregnancies with this notion of gravidity. Interacting with the medical establishment quite literally makes pregnant humans *into* gravid specimens. The violence of that association has been truest for people of color, stretching back to the abuses of enslaved women as research subjects of early obstetrics and lingering into the present with stark racial disparities in maternal and infant mortality rates. At the same time, pregnancy and childbirth—bringing life from shadows into light—are perhaps the wildest, most animal things a person can do.[7]

In the eighteenth century, naturalists feverishly quantified and described every facet of the animal specimens they collected. This was especially true when it came to matters reproductive. In possession of a dead rattlesnake, they would count the number of joints in its rattle, measure the snake's length and girth, tally up its total number of scales, and, were it a gravid female, slice its abdomen to unearth and account for the offspring inside, as when the colonial traveler Jonathan Carver carved open one rattler with a reported (and perhaps exaggerated) "seventy young ones in its belly . . . perfectly formed."[8]

Herpetologists continue to carry out aspects of this calculus. Their practice relies on unfettered manual access to specimens, to bodies. They count up a snake's scales to differentiate species and sexes, for instance—even though it can be gallingly difficult to get right, as seen by Kalm's gripes when issued a similar edict by Linnaeus. For male rattlesnakes, biologists quantify and characterize each minute feature, fringe, spine, and lobe of the rattlesnake phallus, a two-pronged member known as a hemipenis that resembles dewy barbed bunny ears. For females, as shown by Brendan's labors, they surveil pregnancy and postpartum life, enumerate the young. They even track when female snakes are getting ready to mate by showing signs of yolk formation, or vitellogenesis. So writes Ted Levin with more than a hint of excitement in *America's Snake,* a book about timber rattlers and the scientists who study them: "I've run my fingers down the sides of vitellogenic rattlesnakes, felt a hint of their future. Subtle and symmetrical, the contours of prefertilized eggs are the living beads of an ophidian rosary, one after the other, round and promising with yolk." It might be worth noting here that rattlesnake scientists—way back when and now—and rattlesnake writers—historical and contemporary, historian and otherwise—are overwhelmingly male.[9]

Each prenatal checkup, I would lie down, exposed like a curiosity, while a doctor, midwife, or more typically a resident-in-training calculated the size of my growing uterus with a tape measure and palpated my bare stomach. They rarely referred to me by name; I was suddenly "mom," or "she." Medics in masks logged each pound I gained, pondered every blood pressure reading, sought any trace of protein in my urine. I almost thought I fooled them when my apparently non-binary blood type went undiagnosable for months—until they found a term for that, too. Once I started experiencing heart palpitations from the increased blood flow of pregnancy, a cardiologist fastened electrodes across my chest and ribs, which he connected by wires to a device that monitored my internal rhythms for several days, making me feel not unlike a radio-tagged snake.

During labor, I would be strapped with monitors again to chart the *thump thump* of my baby's heartbeat and the crescendos of each contraction. Had I undergone a caesarean section, I would have been sliced open from the belly and my contents extracted like Carver's rattlesnake. Birthing at a university hospital, I was a pedagogical specimen, like many an Enlightenment-era snake. Accommodating, pliant, compliant, and female, my body was a teaching moment, a canvas for heuristic mistakes. (One blunder by an unsupervised resident during labor could have become life-threatening, had I not caught it.) My baby, as well, became an object of study and specimen to track, her early weeks marked by an obsession over pounds, ounces, minutes of milk, inches from head to toe, centimeters of head circumference. When I first saw her on this side of existence—

But there are still things I must keep from you.

"Vulnerability has always attended being pregnant," writes historian Sarah Knott in *Mother Is a Verb,* a memoir-history of motherhood through the ages. "Projections of vulnerability on the visibly pregnant, too," she adds. Both rattlesnake bodies and pregnant human bodies have been viewed through a lens of danger: one capable of inflicting it, the other of receiving. But vulnerability cuts all ways. Historically and now, gravid females have been the most exposed snakes in any timber rattlesnake community, as pregnancy prompts them to bask conspicuously on rocks, heavy with young that distend their bellies and spread their scales. In distinction to a human pregnancy, they will also forgo food during gestation and eventually lose around a third of their body weight, which takes years to regain. Even as females in this condition remain vulnerable, they are also a snake community's cornerstone and perpetuating force. The years-long intervals between each snake's gestation means that, in the words of recent biologists, "Removal of a single animal, especially an adult female . . . has a relatively high negative effect on the population, damaging its ability to sustain itself." Any act of gendered violence pushes this species closer to local extinction.[10]

These snake matriarchs guard the next generation, too. After giving birth, the assembly of basking mothers will stay by the sides of their vulnerable newborns and protect them for a week or longer, until propelled by hunger into the forest for that first, precious postpartum meal. Rattlesnakes have their village. Familial affection and ties of kin in the cold-blooded reptile world have often been hard for scientists to see, though they have always been there.

Take the small carnivorous dinosaur known as *Oviraptor,* whose name translates to "egg thief." Ever since the dinosaur's discovery in the 1920s, when a fossil uncovered in Mongolia showed an adult specimen hovering over a clutch of oblong eggs, paleontologists assumed the dinosaur was scavenging the eggs right when catastrophe hit and preserved them all in a prehistoric snapshot. It took until the 1990s for Mark Norell, after finding an actual oviraptorid embryo in another oblong egg and then a fossil of a dinosaur actively *sitting* on a nest in full brooding position, to fully convince and document for the paleontological community that these reptilian raptors—the kin of modern birds—were parenting just like birds do. In fact, modern birds probably parent *because* the practice existed among ancient dinosaurs, birds simply being those dinosaurs who averted extinction. Some scientists suspect that, contrary to human norms, it was the *Oviraptor* fathers, not the mothers, who sheltered the eggs, given the ubiquity of male parental care among ostriches and cassowaries, the so-called ratites and most basal and dinosaurian of modern bird groups. As with the terrible lizards, it has similarly taken some time, retraining, and mental openness for naturalists to appreciate rattlesnakes not as unemotive killers and loners, but as parents, too: mothering as not just warm and soft, but also cold, scaly, and hard.[11]

A mile into the woods, we were surrounded by female timber rattlesnakes, assured by the *beep beep beep* at every frequency of Brendan's radio. We knew they were there, hidden from view. One, we determined, had found refuge somewhere in a nearly vertical cliff wall, which dropped beneath our toes many dozens of feet. We debated sending a non-gestating person to scout but agreed it would be far too treacherous for anyone to attempt to scale from our vantage at the top. We could take time to circumvent the cliffside and try from the bottom, but there were still no guarantees, and we would be further, and downhill, from other signals, potentially without a snake to show for it. Knowing the terrain, Brendan could tell from the beeping that a second rattlesnake had traveled across a dense marsh that would be similarly difficult to maneuver. A third was just too far. Trying to find these snakes would have been dangerous on this oppressively hot day.

One lone signal, however, held out an electric promise. It hailed from a direction that, though deep in the forest, wasn't filled with cliffs or marshes, according to Brendan. We put our faith in this sound, and in her direction we went, eager to find this subtle serpent, not-so-subtle antenna hoisted overhead.

Along the way, we heard the call of an ovenbird, a warbler with a ruddy crest that builds nests on the forest floor in the shape of a bread oven. So, in addition to watching for untagged rattlesnakes who might be in our path, we also now gingerly worked to avoid smashing any of these delicate ground nests. I morbidly thought of how good a feast those broods of baby ovenbirds would make for postpartum rattlers. After what seemed like an eternity of trying not to step on anything animate, Brendan pointed his antenna at a fallen tree and an adjacent patch of bramble.

And I knew from his body language that we had found her.

We approached slowly, reverentially. She was as unmovable as a mound of brain coral. Due to her infrared-sensing abilities, we must have appeared like lumbering towers of magma. The serpent's gaze almost seemed to narrow at us, an artifact of a rattlesnake's characteristic supraocular scales, which create a permanent hood above each eye. Without the aid of technology, we likely never would have found this statuesque snake curled quietly atop a scrim of dead leaves. She was a lighter charcoal color than most of her kin out here, her hallmark banding pattern faintly visible like veins in marble. Brendan estimated that she was four feet long when fully extended and that she could strike about half the length of her body from this pose. We remained more than a couple of feet back. Brendan recognized her as a postpartum female who had given birth to at least six live young the previous fall. She was near the start of the long journey of a mother's recovery, not to give birth again for several more autumns. The offspring she had sheltered and cared for were now in the woods as well, probably slurping up ovenbird eggs.

Brendan described her as confident. She barely budged when we approached, assured of her own power. Yet, he noted, she probably didn't *want* to resort to expending her venom if it could be avoided, given the cost of cultivating it. I wondered if she may have also been afraid: she was one of the snakes who had been radio-tagged through surgical means. Ironically, as we gathered

Figure 4.1. The timber rattlesnake we found. (Photograph by William Robles)

around this stoic and highly potent being, a commonplace bald-faced hornet began to divebomb our heads, its venom posing more of an immediate threat than that of the calm snake. In that moment, Brendan and I jovially agreed that we'd both rather die by rattlesnake bite than a horde of wasps. William wasn't so sure. Brendan found a long stick and used it to lightly lift the veil of bramble covering the serpent's face to allow us a better photo op (figure 4.1 and plate 6).

She made no sound at the stick's approach, demurring even to reveal her eponymous rattle, let alone shake it in alarm or defense. William said she had the air of a grandmother, putting him at immediate ease. He told me later that he felt a desire to move closer—an urge he suppressed out of respect, both for biologist and snake. I, too, felt a magnetic force, similar to what one eighteenth-century Vermonter named Elias Willard experienced. Initially planning to kill a rattlesnake he encountered, Willard wrote that "my curiosity led me to view him. . . . while I forcibly dragged off my body, my head seemed to be irresistibly drawn to the enchanter, by an invisible power." As much as rattlesnakes ask us to look away, they also beckon us to behold. Enchantment, it seems, is one of the truly unresolved corners of the eighteenth-century study of rattlesnakes. The early modern period saw a much richer body of theorization about possible causes and meanings of enchantment than scientists, and really most historians, would dare seriously entertain today. If herpetologists of the past few decades are just now catching up to early Native observers in their recognition of rattlesnake sociality and kinship, what else might modern science be missing or dismissing from the centuries-past corpus of snake knowledge? There are still things we don't understand about the sway animals hold over us.[12]

Brendan gently lowered the bramble over the serpent's face again, leaving her as obscured as we had found her. And that was that. Our encounter had to be brief, by design. Lingering any longer might have perturbed her. "We don't want to distress endangered species," Brendan said, unironic about the understatement. We said our goodbyes to the snake and began our way back. On our descent, we decided to travel down and around to the base of the cliffside from before to see if we could glimpse the snake whose signal had appeared first on our radar. But peering up from the bottom confirmed our fears: the stone fortress was not safe for human wandering. The rocks formed a theater in the round that scattered our radio waves, making any effort to pinpoint that snake's location a ruse. Just by looking, we could see why the snakes would call this architectural stronghold home. Nothing else in the forest resembled it. I resisted the urge to take a photograph, as that somehow felt disrespectful, or profane. Nor did I want to tempt myself with a record that could lure me into retracing my footsteps.

That we could only find one snake made sense in retrospect. Her singularity was compelling: a reminder of cryptic and unseen worlds underfoot, of invisible scent trails and pheromonal chains linking these snakes in a dispersed network of communion, of gestations best left hidden from human view, and, more solemnly, of vastly diminished snake numbers in this region. I envisioned her, figuratively, as a matrilineal descendant of I.II.I.a, a lone female rattlesnake monitored by scientific men.

We didn't see a single human soul on our adventure. Perhaps for the better. I wondered if this snake had any inkling of the storms of the human world raging around her. If the pandemic retreat of people into homes had coaxed her and the three others to venture farther, wider, wilder than before. In the minds of Brendan and his colleagues, their task wasn't simply a matter of ensuring people don't find the site. They saw victory not only in humans losing some knowledge of rattlesnakes, but in rattlesnakes losing some knowledge of humans, too. For eastern timber rattlers, memory of the human is a tragedy.

"Into the underland," writes Robert Macfarlane, "we have long placed that which we fear and wish to lose, and that which we love and wish to save." Coexistence of humans and rattlesnakes is fundamentally fraught. Yet efforts to completely divorce the worlds of snakes and people are also futile. If snakes can elude human knowledge, they can also elude human attempts at estrangement. One quite conspicuous trace of humanity already encircles the den—though to avoid divulging the site, I once again won't say what. The three biologists who penned the aforementioned diatribe against the snake-slaying Komarek write: "The timber rattlesnake is one of the last symbolic wilderness species remaining in eastern North America. Many of our mountainous deciduous forests would be lacking an element of pristine excitement without this species." Such a vision of wild species and wild lands would have been incomprehensible to settlers who arrived in the New World in the seventeenth and eighteenth centuries. To them, wilderness was a vile and desolate thing. And it was never really wild, given the manifold ways that Native people shaped, and continue to shape, the landscape. The environmental historian William Cronon delivered the most famous and trenchant critique of the concept in 1995, writing that "there is nothing natural about the concept of wilderness. It is entirely a creation of the culture that holds it dear, a product of the very history it seeks to deny." Wilderness, far from preceding humans, in fact depends on civilization for its creation. That most snake biologists I've spoken with don't know that they are in significant ways borrowing a long-standing Native practice likewise points to a collective amnesia of how these lands and animals have been managed since long before Europeans ever set foot on the continent.[13]

But even if the notion of a pristine nature is illusory, something did tangibly shift for these snakes when European settlers invaded. Now, even more changes will afflict them due to further upheavals in our global system. Remove human bodies from the den of the snakes, if you can or should, and the tentacles of climate change can still find their way in.

Looking back into the green hush of the woods, hand on the growing orb of my belly, I wanted to ask that snake a question I had been quietly mulling for months. *How do you mother when the world is burning?*

Plate 1. Isaac Lodewijk la Fargue van Nieuwland, *Portret van Laurens Theodorus Gronovius met Zijn Kinderen,* 1775, watercolor. (Courtesy of the Museum De Lakenhal, Leiden)

Plate 2. A southern cassowary (*Casuarius casuarius johnsonii*) in Far North Queensland. (Photograph by William Robles)

Plate 3. The author over Endeavour Reef. (Photograph by William Robles)

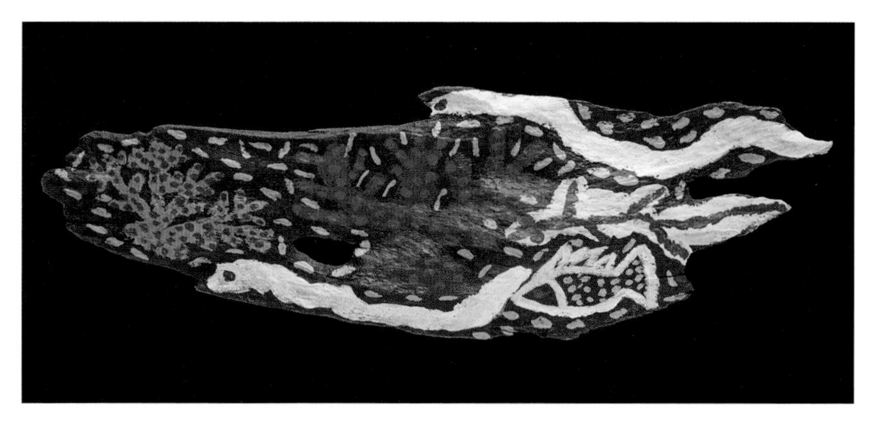

Plate 4. Josie Olbar, *Dulngku* (coral reef), 2019. (National Museum of Australia [MA103108141]; photograph by Jason McCarthy; © Josie Olbar/Copyright Agency; licensed by Artists Rights Society [ARS], New York, 2022)

Plate 5. Gertie Deeral, *Dangerous Reef,* 2019. (National Museum of Australia [MA99250097]; photograph by Jason McCarthy; © Gertie Deeral)

Plate 6. Timber rattlesnake (*Crotalus horridus*). (Photograph by William Robles)

Plate 7. Lumpfish (*Cyclopterus lumpus*). (Photograph from iStock)

Plate 8. William Dandridge Peck, dried specimen of *Cyclopterus lumpus* (lumpfish) on paper, 1793. (MCZ 154782, Ichthyology Department, Museum of Comparative Zoology, Harvard University)

Plate 9. Porkfish (*Anisotremus virginicus*) specimen from 1770. (LINN 147, the Linnaean Fish Collection; by permission of the Linnean Society of London)

Plate 10. Culpeper-type compound microscope by Matthew Loft, with rayskin shagreen covering. (Courtesy of the Collection of Historical Scientific Instruments, Harvard University)

Plate 11. John Cary, pocket globe with sharkskin shagreen case, 1791. (National Library of Australia, 5213930; photograph by William Robles)

Plate 12. Eagle ray in watercolor and pencil by Johann Georg Adam Forster, May 10, 1774. (By permission of the Trustees of the Natural History Museum, London)

Plate 13. Cozumel raccoon. (Photograph by Whitney Barlow Robles)

Plate 14. The author communes with a Cozumel raccoon. (Photograph by William Robles)

FISH

Split the lark - and you'll find the Music -
Bulb after Bulb, in Silver rolled -
Scantily dealt to the Summer Morning
Saved for your Ear, when Lutes be old -

Loose the Flood - you shall find it patent -
Gush after Gush, reserved for you -
Scarlet Experiment! Sceptic Thomas!
Now, do you doubt that your Bird was true?

—*Emily Dickinson*

CHAPTER FIVE

The Flounder and the Ray

ONE YELLOWING FLOUNDER in Harvard University's oldest collection of animal specimens falls flatter than all its flatfish brethren in the sea. In 1793, a self-taught American naturalist named William Dandridge Peck slit the creature in half from head to tail. After removing organs, bones, and flesh with his hands, he firmly pressed one side of the flounder's skin to paper, sewing the animal in place and drying it as if it were no more than a flower plucked from the field (figure 5.1). Annotations in the margins of negative space surrounding the fish list its species name, collection locality, and year of death. Although this specimen, along with many others, has been re-mounted and re-inscribed since that initial exercise, Peck made several dozen flattened half-skins of fish in like manner, diligently cutting, gutting, pressing, and waiting. After graduating from Harvard in 1782 and a brief stint in a counting house, Peck learned how to prepare specimens as he lived and farmed with his recluse father in Kittery, Maine. He went on to become a professor of natural history at Harvard in 1805. Though he was one of the most reputable zoologists in the young United States, he is a little-known figure today.

Peck's practice of unifying animal with page drew on a long tradition of flattening fish, plants, insects, serpents, zoophytes, and other organisms on paper. In the bowels of museums across the United States, Great Britain, Sweden, France, Denmark, Spain, Portugal, and elsewhere, hundreds upon hundreds of paper fish from this era rest stacked up in piles, tucked away from

Figure 5.1. William Dandridge Peck, dried flounder skin mounted on paper, 1793. (Collection of Historical Scientific Instruments, Harvard University, 0038; courtesy of the Collection of Historical Scientific Instruments)

public view. Some call them fish herbaria, given their resemblance to plants preserved on sheets. Specimens stuck to slips of paper built the backbone of major Enlightenment-era collections like that of Linnaeus, Peck's personal hero. Peck's own archive includes a ribbon-thin gunnel, a sturdy carp, and the lumpfish—a colorful, multidimensional wonder in life, as suggested by its common name (figure 5.2 and plate 7). Now, in death, it resembles a deflated balloon, its vivid hues flatlining into a uniform brown (figure 5.3 and plate 8). European collections showcase an even wider range of eighteenth-century forms. In hidden chambers and bomb-proof vaults that insulate their contents from the cold wars above, I have glimpsed the obstinate skins of armored catfish, the top halves of stingrays, collapsed pufferfish, shoeleather shark skins, and lengthy garfish and eels folded over themselves to fit on paper; fish with giant spined sails that could pierce a pinky and others no larger than a thumbprint; specimens strapped in place with a network of strings, fragile seahorses secured with pins, and many more hastily varnished onto the page, as if to become one with it; a sargassum fish, that natural dissembler so brilliantly camouflaged to make like a twig in its seaweed habitat, finally exposed (figure 5.4). Fish from the globe's imagined corners, corralled in a central mausoleum.

Figure 5.2. Lumpfish (*Cyclopterus lumpus*). (Photograph from iStock)

Figure 5.3. William Dandridge Peck, dried specimen of *Cyclopterus lumpus* (lumpfish) on paper, 1793. (MCZ 154782, Ichthyology Department, Museum of Comparative Zoology, Harvard University)

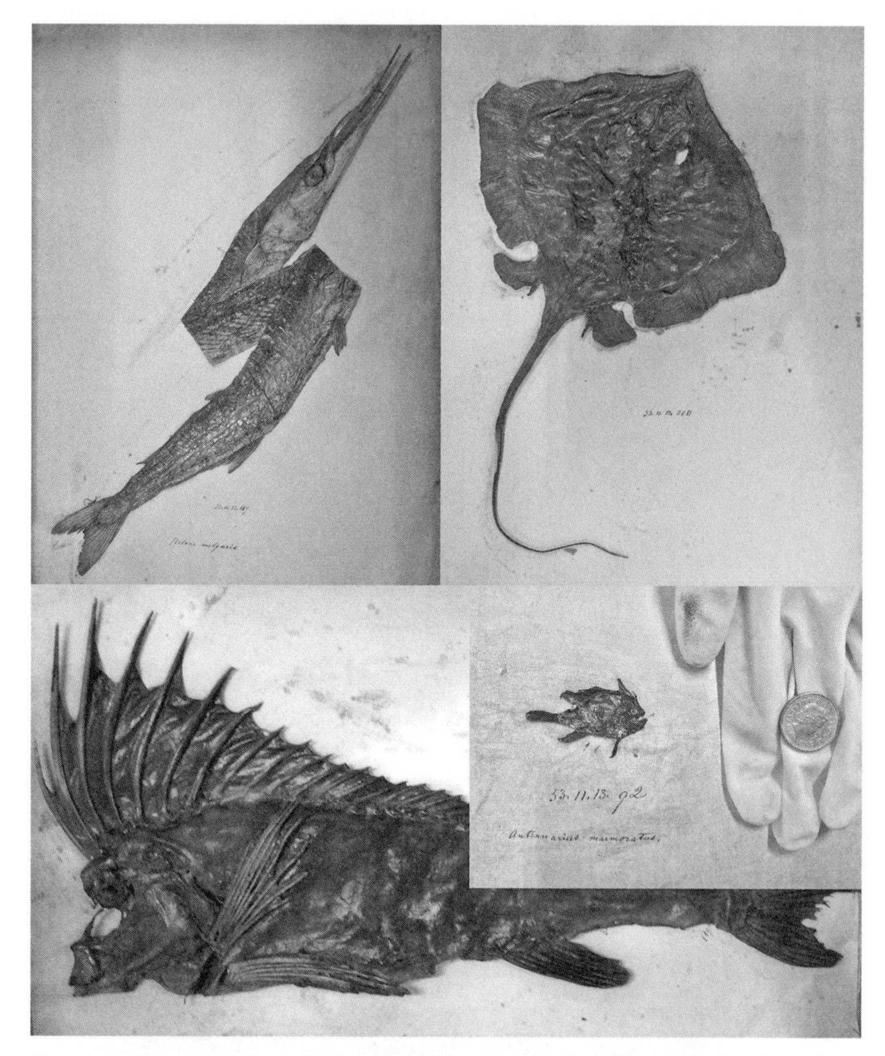

Figure 5.4. A montage of eighteenth-century fish specimens from the Gronovius collection, housed at the Natural History Museum, London. *Clockwise from top left:* a garfish, common stingray, sargassum fish, and horsefish. (Photographs by Whitney Barlow Robles; by permission of the Trustees of the Natural History Museum, London)

These overflowing treasuries, awash with skins and inscriptions, all raise a seemingly simple question: Why flatten a fish?

Eighteenth-century instructions for flattening fish often say precious little about the motives or meanings behind these specimens. But looking at their messy creation and far-reaching use suggests the many roles played by this early form of cut and paste. Flattened specimen preservation developed in an imperial world. At a basic level, it enabled the transit of fish and other animals across the earth in essentially picture form. Each fish became fused with metadata and annotations written on its paper background or even, at times, on its own papery skin. As a form of information management, this fish paperwork made the animals of the far reaches of empire visible and readily accessible. It facilitated a bestial colonialism, achieved through the creation of universal taxonomies. The more animals a naturalist had at his (and it was usually *his*) fingertips, the better he could organize the grand panorama of nature and vie for the title of second Adam. Flattened fish were not merely objects to be shipped efficiently. They were surfaces for making and storing knowledge.[1]

Peck's flounder, and so many other specimens flattened on paper like it, also literalized the pervasive practical metaphor that natural history was an act of reading the "book" or "page" of nature, a grand epic written by God. Indeed, such specimens came to resemble drawings or prints more than they did swimming animals. When forced to become flat, life imitated art. Flesh became word. Flattening natural objects rendered them page-like and canvas-like, making specimens legible as a text and interpretable as objects of natural theology. Some naturalists had their plant and butterfly herbarium sheets bound into volumes, some pasted organisms into existing tomes, and some stored their specimens on bookshelves. Others scribbled directly onto flattened fish skins, underscoring their status as nature's page (figure 5.5).

Is that, then, why they flattened a fish? The full answer is still more complex. For it often had as much to do with the fish as the flattener. We must now turn to how the dead haunt the living: how naturalists tried, sometimes bunglingly, to make animals stable, known, and transmissible on paper, and how animals—even those departed and dumb—had something to do with it. These fish were not merely mute objects bent by human hands. As a living subject, that flounder influenced the conditions by which Peck came to know it in death. During its lifetime, the creature underwent transformations that almost encouraged its preservation on a piece of paper. A larval flounder enters the world, like most fish, with an eye on either side of its head. But as it develops, one eye wanders across the skull to join the other on a single side, while the fish's body compresses to adapt to its seafloor habitat. It transmutes from three dimensions into something more like two. As bottom feeders and bottom dwellers, flounders are flatness par excellence. They make themselves especially available for a paper eternity.

Figure 5.5. The inscribed skin of an eighteenth-century smooth puffer (*Lagocephalus laevigatus*) specimen in the Gronovius collection, housed at the Natural History Museum, London. (Photograph by Whitney Barlow Robles; by permission of the Trustees of the Natural History Museum, London)

Such a creaturely quality mattered for natural history. In the words of theorist Bruno Latour, "Scientists start seeing something once they stop looking at nature and look exclusively and obsessively at prints and flat inscriptions." During the eighteenth century, natural history was just as engaged with visual and material representations of nature as with words on a page or lived field experience with animals. Drawing, painting, diagramming, and printing strove to inscribe fleeting life into semi-permanence on two-dimensional surfaces. In the case of fish herbarium specimens, the visual and material converged. Flatness was simultaneously a method of specimen preservation, a way of animal being, and a philosophy of seeing.[2]

These were not bare or straightforward transcriptions of nature as it really was, however. Visuals made their own meaning. Far from granting a transparent window on nature, flattening entails loss by definition: at minimum, any translation from three to two dimensions produces some form of distortion, to the endless chagrin of cartographers. In Latour's words, the flat inscriptions of early modern science had, in comparison to their three-dimensional counterparts, "been *made less confusing*." Many eighteenth-century naturalists embraced this flat-but-incomplete visual language of natural history. Mark Catesby nodded to flatness as a category of representation when telling his readers: "I humbly conceive Plants, and other Things done in a Flat, tho' exact manner, may serve the Purpose of Natural History, better in some Measure than in a more bold and Painter like Way." But natural history's two-dimensional approach also entailed a particular politics of seeing and making. Flattening squeezed out rich social and environmental contexts behind the creation of knowledge. It excised the contributions of marginalized actors like enslaved taxidermists, auxiliary practices like cooking, and the essential domestic work of women. For all we know, perhaps it was not even Peck, or not only Peck, who made Peck's specimens; no evidence rules out a helper having lent an expert hand in sewing or fish-filleting. Flatness, as we will learn, proved deft at covering its tracks.[3]

Even as many naturalists employed flattened methods in hopes of standardizing nature and making it stable and visible, the lifeways and material form of fish and other creatures shaped, and at times confounded, such efforts. Fish themselves enabled and eluded naturalists' attempts to collapse the natural world into two dimensions: not even all that was flat translated so easily to paper. Like a flounder, stingrays, skates, and other so-called batoids have compressed body structures and tend to live on sandy bottoms, where flatness helps them blend in with the seabed. Certain ray and flounder species form part of the same seafloor ecology. Yet unlike Peck's flounder, rays *lose* many of their features when preserved on paper due to their top-and-bottom anatomy. They make an awkward fit for the fish herbarium, specifically by foiling the profile view. By maintaining a kind of twist and "torque," to borrow from anthropologist Christopher Pinney, rays forced naturalists to literally pivot, evading the simpler convergence of animal and illustration—of nature and book—that many other fish permitted. Rays thus function in this story as both historically real creatures and as a principle of the limits of flatness in natural history's visualization project. At the same time, rays, as we will see, generated alternative methods of knowing: a textural rather than textual mode of doing science.[4]

Despite Enlightenment science's stated research program of rationality, classification, and visibility—traits that on their surface seem to be shared by flattened specimens—animals like rays reinforced what was unknowable, or visceral, or unreasonable and unreasoning. Flattened fish aided the work of building universal systems and acquiring total knowledge of nature, but they also undercut it. Animals escaped naturalists' scrutiny, partly due to mundane facts such as what could or could not be caught, or what would or would not fit on paper, and partly due to the profound difficulties of ever comprehending nonhuman life. Flat specimens bore distortions and omissions that betrayed their inability to stand in for the original, casting doubt on the metaphor of nature as text or image. As we see by looking at collections, then the collectors, and finally a return to the collected, fish's enduring slipperiness to human probing and certain forms of archiving prompted naturalists to acknowledge gaps and fractures in the natural historical project.

Do they not escape us still?

COLLECTIONS: FLATTENING NATURAL HISTORY

Most biologists today regard "fish" (or more technically "fishes" when referring to two or more different species) as an unnatural grouping. Something like a flounder and a ray bear only very distant kinship. Nevertheless, naturalists approached this category of animals through a shared lens in the past. Fish sat in a middling station on the Chain of Being, neither low and imperfect like a polyp nor intricate and human-adjacent like a raccoon or ape. Fish do stand out, however, as some of the more unusual candidates for paper preservation given their watery origins. They also stand out as the most numerous flattened animal specimens that are extant from this era, even though naturalists also turned to flattening butterflies, snakes, corallines, and other organisms (figure 5.6).[5]

"The Sea affords an inexhaustible field for the researches of the naturalist," Peck, son of a shipbuilder, wrote to the Reverend Jeremy Belknap of New Hampshire in 1790. "The discoveries which have as yet been made in it, are comparatively nothing." Although people had been voyaging across seas for centuries at the time of Peck's writing, the many thousands of species that commanded the waters beneath their ship hulls were rife with mysteries. What we know of marine life today is a mere hint of what remains to be fathomed. But by the late eighteenth and early nineteenth century, Americans in their new republic felt particularly suited to aid the progress of ichthyology—the study of

Figure 5.6. Dried and pressed coralline specimen on paper, c. 1767. (GB-110/LM/PF/SPE/3; by permission of the Linnean Society of London)

fish—through offering specifically American contributions to the science. Benjamin Smith Barton, a frequent correspondent of Peck's, noted in 1807 that the natural history of fish had been "extremely neglected in the United-States," especially in regard to fish instincts, manners, and habits—what Barton deemed "the most interesting part of animal natural history." He conceded that their watery world prevented "an easy or a rapid acquaintance with the *mores* of the fishes," but Barton nevertheless faulted the "incurious supineness" of naturalists who painted fish as a "stupid race of beings." Barton's own research had shown him the "ingenuity" and familial "love" of fish, convincing him "that we have detruded the fishes to too low a station in the scale of animal intelligence." He urged his fellow naturalists to study fish and thereby devote themselves to "piscatory pleasures."[6]

The neglect of fish runs as a theme throughout the history of Atlantic science. Whereas the dazzling intricacy and ambiguity of corals prompted heated debates over the boundaries of animal life, and rattlesnakes generated tremendous interest from New World travelers as a singular and symbolic threat, most fish were, by contrast, ordinary. Alexander Garden, who eagerly supplied Linnaeus with specimens from the Americas, admitted to the Swede: "Before the receipt of your letter, I had scarcely paid any attention to our fishes." The English naturalist William Swainson theorized that the "impossibility of preserving the beautiful but evanescent colours of fish, and the unsightly appearance they generally present" when preserved might explain their disregard by collectors and naturalists. Yet fish were also ubiquitous, in part because they were a familiar, even mundane sight on a plate. In an early form of crowdsourcing, Samuel Latham Mitchill recommended seeking dried and flattened fish specimens from the public given the potential for distinctly American scientific contributions to this branch of natural history, not to mention the economic value of even dull-looking fish. Flattening fish could be both an act of citizen science and an expression of patriotism.[7]

Peck noted in his Harvard lectures on ichthyology that the science remained "in great obscurity" until the groundbreaking work of the Swedish naturalist Peter Artedi, often called the father of modern ichthyology. Peck's characterization was a tad sweeping; other major works on fish predated the eighteenth century, such as *De Historia Piscium* (1686) by Francis Willughby and John Ray (although that lavishly illustrated title, financed by the Royal Society, sold exceptionally poorly and sent the institution into somewhat of a financial tailspin). Nevertheless, Artedi's approach was a watershed. Artedi mysteriously drowned in 1735, and Linnaeus, a close friend, posthumously published Artedi's *Ichthyologia* in 1738. The late naturalist's work informed Linnaeus's own organization of fish in his epic *Systema Naturae*.[8]

Peck took more than a passing interest in Linnaeus's (and thus Artedi's) methods. While developing his zoological skills in relative isolation a decade before his Harvard career, Peck expressed a cautious love for natural history, which he attributed to Linnaeus, when he wrote to Barton in 1794: "Put me right with freedom when you find me erring, which it is likely, you frequently will, for it is but a few years since first, *Linnai captus amore,* I have studied nature. I proceed with doubt & deliberation, am desirous to become a naturalist, but would not be a virtuoso." By 1807, once Peck had assumed his Harvard professorship and was traveling Europe to court the great naturalists of the day and gather resources for Harvard's new botanical garden, he confided his intense affection for Linnaeus in a letter to James Edward Smith, founder of the Linnean Society of London. "Did you ever observe," asked Peck with a note of rapture, "in reading any thing which affected the mind by its energy or sublimity, that the blood flowed more copiously to the heart, leaving a chill on the surface of the body? I cannot explain it; but I never read Linné's Introduction to the Syst. Nat. [*Systema Naturae*] without this sensation." (Surely no one has paid it that high a compliment since.) In Peck's possibly apocryphal origin story as a naturalist, he reportedly decided to pursue the study of nature when he found a tattered copy of Linnaeus's tome washed ashore from a shipwreck.[9]

Peck's reverence for Linnaeus translated to his scientific work, as his flattened fish skins specifically lent themselves well to Linnaean-style classification. Instead of salvaging the fish wholesale or making it appear lifelike, Peck preserved and exposed observable diagnostic features needed for sorting the fish according to Linnaeus's system and for teaching this classification to students. Linnaeus often delineated species by using discrete, usually externally visible traits or "characters," many of which could be quantified. To place plants in the tree of life, Linnaeus counted and compared stamens and pistils, the reproductive organs of flowers; to classify snakes, he counted the scales along their curving bodies, causing inordinate frustration among his disciples like Pehr Kalm; and to guide genus and species designations in ichthyology, Linnaeus's system considered various exterior traits, paying special attention to the number and position of a fish's fins as well as the quantity of rays or spines each fin contained. Eighteenth-century naturalists often referred to fish simply as "the finny tribe." Counts of the number of rays per fin can be found on the paper substrate of some of Peck's fish specimens and appear in the text of his publication "Description of Four Remarkable Fishes; Taken near the Piscataqua," regarded as the first article produced by an American systematic zoologist. In his correspondence with other naturalists, Peck often exclusively used "the form and disposition of the fins and the relative number of their rays" as a guide for determining a fish's classification.[10]

Although several sets of instructions for flattening fish exist in manuscript and printed form, Peck likely followed the canonical recipe written by Jan Frederik Gronovius and published in the Royal Society's *Philosophical Transactions* in 1742. Gronovius, who also funded Linnaeus's work, appears to have been the first to publish this technique. However, he told John Bartram that the German naturalist Johann Ernst Hebenstreit, who explored parts of Africa for Augustus II in the 1730s, "invented this methode" but "never would communicate the way to prepare them so; till at last I found it out a few years ago." The process involved cutting a fish in half from head to tail with scissors to preserve one half of the skin and all the fins on one side, removing bones and internal organs, and positioning the skin on a board to dry in the sun. After some time, one could then separate the putrefied skin from the flesh, further flatten the skin, and apply it to parchment or paper. Gronovius specifically instructed preparators to spread and pin open the animal's fins during the drying step to prevent these appendages from clumping into uncountable rigid masses—and, indeed, some specimens retain visible pinholes in their fins to this day.[11]

Peck, then, was not alone in his devotion to fins. When Alexander Garden prepared half-skins to send across the Atlantic to Linnaeus, he explicitly connected this flattened mode of preservation with the display of fins, writing: "I have caused their skins to be dried, by which I think you will be able to see the true situations of the fins. This will be more satisfactory to you than a bare, and perhaps inaccurate, description of mine." John Bartram also sent Gronovius "the skin of the Fish, with its fins curiously displayed on paper." The spreading of fins in a dry specimen, which made the animal's rays readily countable, was one of the standout features of these flattened preparations, built into the preservation process itself. It laid bare the animals for human eyes and let fish flesh become numerical; Peck's past vocation in accounting proved most relevant here. Peck additionally used a varnish to affix labels with binomial species names directly onto the sides of some of his specimens. For his own part, Garden numbered the fish he sent to Europe by writing on their skins directly, repurposing them as writing tablets.[12]

These objects thus served the gods of naming and classification, bearing literal annotations in the mode of natural history illustration. As such, flattened specimens helped naturalists—especially metropolitan naturalists—manage the influx of animal information arriving from global waters. Flatness also helped achieve a level of permanence still visible today. Fish prepared in the dried and flattened Gronovius style have outlived many other specimen formats from the era. Two major repositories of dried fish herbaria now housed in London—the collection of the Gronovius family, held at the Natural History Museum, and the collection of Linnaeus, held at the Linnean Society—showcase saltwater and freshwater animals from across the world. Many of these fish have been retroactively classified as lauded "type specimens," those specimens that in essence define a species by serving as the basis for the first published descriptions, even though that concept would not formally exist until the nineteenth century. (This presents its own ironies, as many of the type specimens biologists must now consult as authoritative representatives are only half-specimens.) Each of Linnaeus's flattened fish skins has its own shadow box for safekeeping. He received many of the New World specimens from Garden, and he sent his devotees on collecting missions elsewhere throughout the globe, including Peter Forsskål, whose fish herbarium from the Danish Arabia expedition (1761–1767) now resides in the Natural History Museum of Denmark. Gronovius passed along his flattened fish collection to his ichthyologist son, Laurens Theodoor Gronovius, who authored the treatise *Museum Ichthyologicum*. Laurens continued to expand the collection, and its fish now live tidily stacked in a series of cardboard boxes in the Natural History Museum's basement storage.[13]

When naturalists chose not to dry fish specimens, they usually submerged the animals whole in alcohol ranging from brandy to wine. Some naturalists preferred this method since one could retrieve the fish to peer inside the gills and mouth while also pulling fins apart to count the number of rays. But inspecting fins in this manner required dirtying one's hands. Preserving fish in alcohol presented practical challenges as well, since liquor could evaporate during transoceanic journeys and specimens might spoil if they touched the bottom of the bottle or if jars became overstuffed with specimens, as often happened when space or containers were limited. The shape a fish's afterlife took often boiled down to logistics.[14]

Specimens in spirits were also simply much bulkier to ship than sheet-like fish layered efficiently in a box. Recall those instructions for preparing natural history specimens to ship overseas from various colonies to Europe drawn up by the London-based collector George Humphrey, who also dunked corals in liquor. Humphrey's handwritten 1776 booklet provides a rare glimpse into the reasoning behind the Gronovius-style technique. "By this Method," Humphrey wrote, "many kinds of Fishes may be so preserved as to take up very little more room than a Drawing, and will look infinitely better." Humphrey's comment explicitly shows the illustration-like nature of these specimens while also hinting that compression facilitated their journeys across seas or into storage elsewhere. Drying and flattening fish did away with the messiness of retrieving animals suspended in spirits. The number of rays per fin were spread out in plain view for the onlooker, like a picture, and often also written out in ink, rather than being folded against one another and requiring manual manipulation of a slippery pickled specimen. Flattening also drew on the visual language of natural history illustrations and book plates by isolating animal bodies on blank backgrounds—a mode of decontextualization that the art historian Janice Neri has called "specimen logic." The margins of some fish herbarium specimens feature floating fragments and cutouts like fins or cross-sections, another common trope of natural history illustration. Preparing a fish like the page of a book made visible a being that existed in shadowy, murky depths. It removed the fish from its watery milieu and brought it into ours.[15]

Many of the naturalists who created lasting fish herbarium collections were foremost botanists by trade, including Peck, which further underscores the textual and visual nature of these objects. Naturalists almost invariably identified fish half-skins as plant-like. Without preexisting botanical traditions, Peck's specimens might never have existed. In his autobiography, Linnaeus noted the resemblance of his own fish half-skins to botanical specimens, writing that he stored "in his cupboards innumerable fish glued on paper as if they were plants." One commentator noted this method of preservation made the fish "like specimens of a *hortus siccus*," a frequently used Latin term meaning "dry garden" that referred to a long tradition of drying plants and preserving them two-dimensionally in herbaria. Gronovius told Bartram that he sent "dryed fishes, to be kept as plants in an Herbarius." Also in a letter to Bartram, Peter Collinson mentioned a dried fish collection of Bartram's son William, writing: "I am pleased to see that he has gott so pretty a Way of Drying Fish. Bye it we may have a Hortus Siccus or rather Oceanus Siccuss of Fish."[16]

In their print-like or book-like format and usage, herbaria quite consciously embodied the metaphor of the book or text of nature, serving, in the minds of naturalists, to illustrate God's authorship and artistry. Herbaria converted the book of nature that was out in the world into a tangible library. The notion of nature as a text had existed since antiquity, and it gained momentum with the spread of printing and Protestantism. "There are two bookes," wrote Thomas Browne in his 1642/1643 *Religio Medici,* "from whence I collect my Divinity; besides that written one of God, another of his servant Nature, that universall and publik Manuscript, that lies expans'd unto the eyes of all." The rise of natural theology and so-called physicotheology in the seventeenth and eighteenth centuries provided an empirical framework for understanding God's providence and the regularity of his laws. Naturalists believed one could read signs of craftsmanship throughout the natural world, treating fish skins as a form of scripture, as patterns of a divine mind.[17]

In delivering a sermon dedicated to Peck on the Sunday after the naturalist's death in 1822, the Reverend John Thornton Kirkland, president of Harvard at the time, said Peck "was accustomed to see God in his works." Peck proclaimed in one of his lectures as a professor that "every Tree, every herb is a Volume" of the Creator's miracles. He relied on his flattened fish specimens to guide students through the intricacies of nature's book. Peck's specimens appear to have been stored in the University Museum on the second floor of a building known as Harvard Hall. The museum formed part of a teaching and exhibition space that included a library, a scientific apparatus closet, specimens of natural history and other collections, and a resplendent lecture room covered in red flocked wallpaper called the Philosophy Chamber. It is likely that Peck delivered many of his natural history lectures here, though some were also presented in Boston and at the college's new botanic garden, of which Peck was designer and curator.[18]

It was also in the context of these teaching spaces, these theaters of the book of nature, that Peck's flattening method threatened to, quite literally, break down. Peck stressed the visual rather than tactile nature of flattened specimens when he cautioned his pupils while displaying one fish in a lecture: "The scales are very loosely attached, & fall with a touch tho even so gently handled." Dried fish were legible to the eyes and intellect, but not necessarily to the hands. Their own mode of preservation could also be their undoing.[19]

In addition to keeping students from dislodging the scales of his specimens,

Peck desperately sought to restore color to their brackish bodies through thick descriptions of the live animals in his lectures. Colors often vanish quickly from a fish drained of life. Gronovius, too, rued in a letter to Bartram: "The great misfortune is, that the colour perish; else it shows a good way to find out their characters . . . by the nummer [*sic*] and position of the Fins, and the bones in them." Try as they might to strong-arm nature onto a canvas, naturalists faced paper specimens quite changed from their counterparts in the sea, evacuated of vital features like hue and depth, and beset with distortions, shrinkages, and skewed proportions. Peck, for instance, wrote at length about a compressed fish specimen described by Johann Reinhold Forster that gave rise to unnatural tubercles and protuberances after desiccation. Drying and confining the bulbous, the jutting, and the three-dimensional to a flat surface necessarily warped the final product, like the far reaches of Mercator-projected maps.[20]

Indeed, animals were hardly the only phenomena that underwent the robberies of flattening in the eighteenth century. Remember: scientists see things better on paper. Two-dimensional renderings proliferated across media, evident in maps and charts, lavish natural history illustrations, and the transport of other environmental data in paper formats such as sketches, equations, and tables. Other flat or flattened objects entered Harvard Hall's early museum and shared a storage, display, and teaching space with Peck's fish before the collection was disbanded in 1820. There were paintings, engravings, and mezzotints that made flat planes into scenes. Fossilized animals compressed by time. A rattlesnake skin donated in 1778, the thick of revolution, by a "Miss Meriam." Even Peck's own drawings of skulls, showing supposed human racial types.[21]

And then, among all the two-dimensional inventories of this continually expanding and contracting collection, there is one horrifying line item dropped nonchalantly among other line items. A New York merchant named Francis Goelet described the contents of Harvard Hall during a 1750 visit. Before enjoying a glass of wine at the college, he saw jumbled amid other curiosities a piece of "Neegro's hide tand." Later, on a cold January night in 1764, an unattended fireplace in the building started a blaze that would annihilate these human remains, along with many other human and animal relics in the collection.[22]

Goelet's callous description casually equated the body of a person deemed chattel by law with the dried skins of cattle. People, as well, became flat specimens in the long eighteenth century. It was usually without their consent. The collection context of Peck's fish specimens, as with many curiosity cabinets of the day, articulated social hierarchies through appeals to nature, precisely by mixing the study of people and fish and fossils and snakes in a single collection. Innocent as flattened fish may seem, animal specimens never existed in isolation. Linnaeus, quite controversially, listed humans as just another species carved up into varieties in his *Systema Naturae.* But naturalists were ever more likely to convert into specimens those people they placed lower in the hierarchy of being, those who they hoped might insulate white humanity from its animal nature.[23]

And natural history, we will see, relied on the very people it so brutalized to make the world flat.

COLLECTORS: CRAFT AND CONFLICT AT WATER'S EDGE

European naturalists such as Linnaeus and Gronovius, working from urban repositories, trusted dispersed global networks of collectors and informants—such as the American-based Bartram and Garden—to supply specimens, illustrations, and descriptions of distant wildlife. Many of these colonial contributors, in turn, depended on additional local collectors to perform the physical work of extracting fish from water, cutting and gutting their bodies, and translating their skins to paper. Yet few traces remain of these encounters between living fish and unnamed people. Flat specimens blend so seamlessly with the page that their creation seems effortless, obscuring the many people who had a hand in generating each specimen. Minimalist metadata seldom honor the original collectors. Later curators separated some animals from their original paper surfaces, sending valuable annotations to the slag heap of history. And specimens migrated among collections and collectors over time, bringing a little more amnesia with each movement. Every flattened fish has a human story behind it. Only a few are possible to reconstruct.

Natural history hardly had a monopoly on fish or fish knowledge in the eighteenth-century world. Fish shaped and were shaped by commercial exchanges, foodways, leisure, the fisheries of coastal towns, and the expertise of women, Indigenous groups, and enslaved individuals in addition to naturalists. Flat specimens embody the diverse ways people interacted with aquatic creatures—and those modes of human interaction could undergird, advance, or even thwart natural history efforts. Though documentary gaps loom, fugitive glimpses of unsung labor suggest the larger patterns by which animals made their way into more elite naturalists' hands.[24]

As a matter of course, naturalists routinely sent others to collect specimens on their behalf. Alexander Garden, for instance, described the practice in a letter to Linnaeus in 1771:

> To procure . . . natural curiosities, I sent a black servant last summer to the island of Providence. During his stay there, he collected and preserved some fishes amongst other things; but, meeting with tempestuous weather in his return, and being, for several days together, in dread of immediate shipwreck, he neglected all his specimens, many of which perished. Some were fit only to be thrown away, and others were greatly damaged. What remain, such as they are, I shall, by this opportunity, send for your examination. Some fishes among them, whether found in our sea, or in that of the Bahama islands, you may perhaps find to be new.

This unnamed "black servant" was likely enslaved, working at the behest and threatened violence of a master. A slaveowner himself, Garden encountered and treated many other enslaved people while working as a physician in Charleston. He offered this anecdote, and this individual, as a convenient scapegoat to explain to Linnaeus why he sent a mere fourteen flattened fish in this shipment. Writing from dry land's safe embrace, it was easy for Garden to decry the "neglect" of the specimens. He didn't have to choose between saving his own life from a storm at sea and shepherding yet more dead fish into an overseer's hands.[25]

Yet Garden's cold move to preserve his credibility unintentionally gives us a means of tracking how he actually obtained his specimens and thus the labor behind Linnaeus's monumental storehouse. Some of the half-skins this collector salvaged from the hazardous marine excursion still survive. They now live in a basement vault in the Linnean Society of London, along with Linnaeus's other working specimens. One such specimen preserved by the man is a dried porkfish (*Anisotremus virginicus*) flattened on paper, its banded pattern faintly visible across the centuries if one knows to look for it, though the fish's neon yellow dimmed long ago to an archive brown (figure 5.7 and plate 9). Since fish skin can double as a durable writing surface, the handwritten number 10 on the animal's side let me collate the mute fish with the anonymous collector, as Garden luckily numbered the specimens in his shipment to Linnaeus and included those numbers in the letter cited above. Flattened fish skin itself facilitated the persistence of labor so often submerged. Despite Garden's dis-

Figure 5.7. Porkfish (*Anisotremus virginicus*) specimen preserved in 1770 by a possibly enslaved collector for Carl Linnaeus, by way of Alexander Garden. (LINN 147, the Linnaean Fish Collection; by permission of the Linnean Society of London)

paraging remarks, this person's specimens outlasted many others from the period—a testament to his skill and the doggedness of his toil in one of the most important collections in the history of science. Dry specimens from this period even retain genetic material that can be extracted for DNA analysis today.[26]

If, as is so often the case, marginalized labor is most visible in the archival record when things go wrong, how many successful and thus *uncredited* collecting trips might this man and others have made on Garden's behalf?[27]

Beyond forcing collectors into servitude, naturalists also looked to the practice of fishing for help in acquiring specimens. They visited fish markets to find new species, as ichthyologists continue to do today. They forged relationships with fishers. They hopped into boats with them. When Peck complained that too few local fish were "systematically noticed" because "our fishermen are very inattentive to any but such as are esteemed fit for food," he in fact showed the profound reliance of scientific knowledge on the attention of lay food providers—experts who spent their lives learning the habits of fish.[28]

Peck apparently developed a local reputation as a person in need of

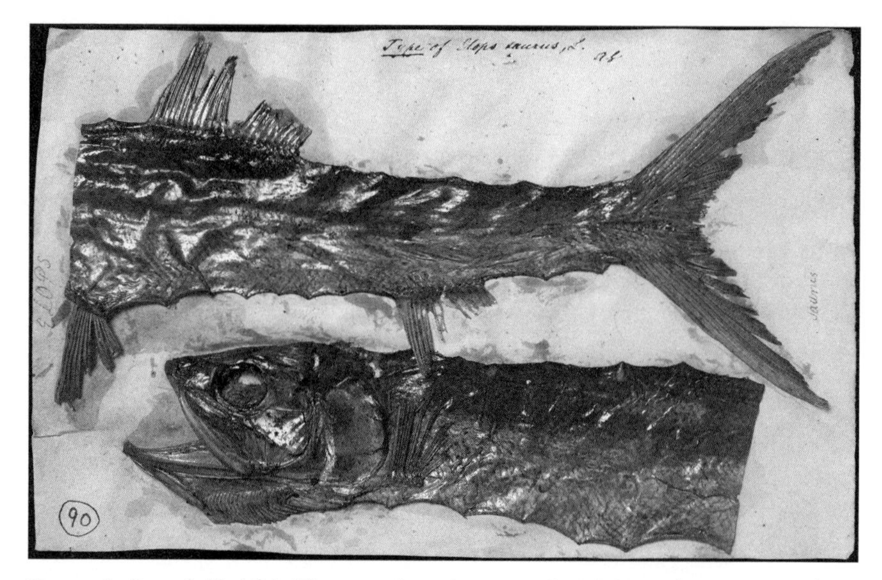

Figure 5.8. De-scaled ladyfish (*Elops saurus*) specimen, c. 1763. (LINN 90, the Linnaean Fish Collection; by permission of the Linnean Society of London)

specimens. He noted that one of his dried fish "was brought to me many years since by a Piscataqua fisherman, who called it a Bream," though he carped that its color was "lost, as it was taken at Sea and was thrown into brine to preserve it," after which Peck or a helper presumably transformed it into a dried skin. Even children contributed to Peck's collection. In his article on New England fish, Peck commented that one specimen "was brought me by a boy who called it a white eel." Although naturalists tended to distance themselves from the knowledge of laymen, manual laborers, women, children, and the enslaved, their accounts nonetheless reveal the persistent influence of extra-scientific human worlds on natural history.[29]

Perhaps the most important role fish played in eighteenth-century life was at the table. The kitchen, like the angler, left breadcrumbs all over the naturalist's repository. Once a fish became a scientific specimen, its edible nature might still be hard to ignore. A flattened type specimen of the ladyfish (*Elops saurus*), also held in the Linnean Society's specimen vault, offers one peculiar culinary story (figure 5.8). Garden told Linnaeus it was "the only specimen of this fish that I ever saw, and the gentleman who was kind enough to let me have it, had unluckily ordered it to be dressed for supper, so that the scales were taken off before he thought of me; and hence you will observe that I could not see the natural appearance of the fish, nor make the characters complete."[30]

Reading between the lines of Garden's phrasing here—the man "ordered" the fish to be prepared for supper, and a South Carolina "gentleman" in the eighteenth century usually had someone else do his cooking—raises the possibility that this type specimen was initially prepared and de-scaled by some sort of domestic laborer. In Charleston, a major hub of the Atlantic slave trade, this very well may have been an enslaved man or woman. Split into fillets to fit on paper, and especially shiny due to its missing scales, the ladyfish embodies a seemingly unscientific relationship between humans and fish that nonetheless left its mark on a natural history repository. As the historian of science Lorraine Daston has shown, type specimens are odd entities. They are "the last court of appeal in all questions and disputes about species definition," even if the specimen turns out to be a wholly unrepresentative representative of a species. This ladyfish makes for an atypical type specimen indeed, as scales were one of those quantifiable features required by Linnaean description. (One edition of Linnaeus's work notes of the ladyfish: "*body* slender, covered with large angular scales.") A type specimen is not the Platonic ideal of a species but, elementally, a hungry fish: the one who took the bait. In this case, a specific moment of contact between a hungry fish and some hungry people outside the collection secured the animal's vaunted status. Any ladyfish specialist today must consult what is left, warts and all.[31]

If eating, or almost eating, could thus lead to knowing, the reverse was also true: many specimens never made it into academic collections at all precisely because they were consumed. Natural historians competed in a broader multispecies struggle for protein. Science sought to preserve fleshy, nutritive bodies that might otherwise enter stomachs. Specimens could travel to collections or the table, but not both. Fish were "boundary objects," to use a term from Susan Leigh Star and James R. Griesemer. They served different and sometimes conflicting ends in the kitchen and the curiosity cabinet, and different needs in different human communities—not to mention nonhuman ones.[32]

Cats, for example, presented a constant threat to the goods of a fishmonger's stall, as portraits of historical fish markets with looming felines attest. Garden lost many of his fish herbarium specimens to "vermin" and to people, as when he griped to Linnaeus about a certain rockfish: "I never had but only one to examine, and the company who permitted me to make out the description, insisted on their having the pleasure of eating it, otherwise I would have preserved the specimen for you." (Garden's correspondence reads as a comedy of specimen errors.) Some of the tastiest fish likely never made it to a scientific examiner. At times, naturalists even consumed their specimens, in whole or in part, effectively incorporating their objects of study into their own bodies. The demands of the hungry mind had to be balanced against those of the hungry body.[33]

This battle against appetite reached indoors as well. Naturalists waged constant war with insects that threatened their specimens, which Peck called "herbarium pests." Overseeing collections involved not only the more dramatic events of describing new species and arranging the natural world, but also ordinary—sometimes verging on extraordinary—acts of maintenance and upkeep. In 1795, the Massachusetts Historical Society printed the following tips, supplied by Peck and drawn from numerous authors, for protecting animal collections from insects:

> These [pests] are all nocturnal insects, and begin to move soon after twilight in quest of proper substances on which to deposit their eggs. The evening is therefore a fit time to examine the walls, by which attention, many of them may be destroyed. . . . The specimens themselves should be frequently and carefully examined, to discover any insects which may have crept into them; without this care, no application whatever will I believe effectually preserve them.

Thanks to live animals, managing archives of dead ones required active and prolonged vigilance and care, even into the night. Sarah Bowdich Lee similarly advocated sustained contact with collections. An English zoologist and explorer, she worked alongside Georges Cuvier and authored numerous books on fish, other animals, and taxidermy in the early nineteenth century (sometimes we must expand our chronology to find more women). Lee believed all herbaria "should be constantly overlooked, and opened, to prevent the destruction caused by insects." This "incessant attention," she insisted, would ward off moths better than any applied treatment. Another collecting guide had a more violent recommendation: check all specimens monthly, and should you suspect one to be insect-laden, "beat it with a stick."[34]

Such closeness and ceaseless care could have a darker side. Peck, who died at the age of fifty-nine, suffered from strokes and paralysis and warned students of his health problems with an air of fatalism, as recorded in scribbles in his surviving lecture notes. Ever resourceful, he even built himself a wheelchair. Ailments plagued Peck as he conducted his research, as well. While in the throes of one extended study of botanical herbarium specimens, which were often generously treated with camphor as an insect repellent, Peck said he gave himself "not more than fifty minutes per day for taking my food," leaving him in a camphorated atmosphere for most hours of the day. Due to the resulting "sensible effect" on Peck's well-being, he had to abandon the project until "exercise & the open air had brought my system to the standard of health." The demands of specimen upkeep strained Peck's already poor state—the man himself an animal body bedeviled by the bodies under his care.[35]

COLLECTED: THE FLOUNDER AND THE RAY

Without the lives and labors of fish themselves, no specimen, no fish soup, no drawing from life could exist. They provided the physical scaffolding and raw materials upon which Enlightenment-era natural history, collecting, and visualization relied. As essential as fish were, they rarely made a naturalist's job simple. Practitioners brought to these beings certain intellectual frameworks informed by natural theology, the book of nature, a hierarchical Chain of Being, and a desire for taxonomic order. The habits, lifeways, and material form of fish could facilitate naturalists' designs, but they also put their own pressure on these methods and metaphors.

Certain animals eluded the fish herbarium altogether or presented other representational challenges. Many sharks and pelagic rays, for instance, proved too large to fit on paper; only juveniles or smaller relatives could be circumscribed in the naturalist's library. George Humphrey noted that fish flattening worked best for "small Fish, that will admit of being skinned, particularly as have large, or hard scales." Eels and other elongated species that stretched beyond the margins of a page forced naturalists to fold the animals back onto themselves should they desire to make them into specimens or printed images (figure 5.9). Indeed, in his *Book of Eels,* author Patrik Svensson dubs the European eel natural history's most enigmatic creature from Aristotle to the present. The eel can be seen clear as day on paper. Yet no person has witnessed the species spawn in the wild or found an egg where they are presumed to hatch, way out in the Sargasso Sea. Paper fails to solve the problem of medium:

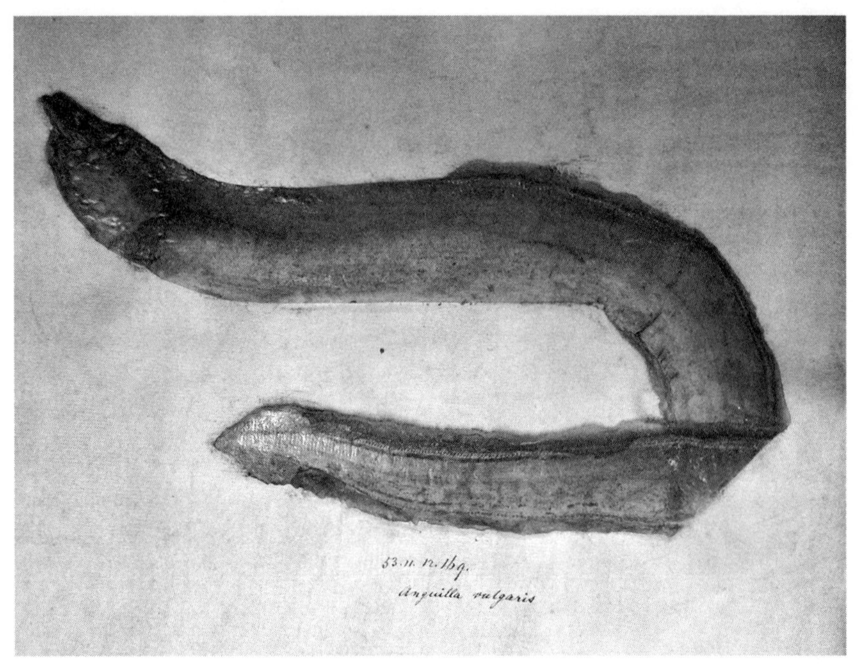

Figure 5.9. A European eel (*Anguilla anguilla*) from the Gronovius collection at the Natural History Museum, London. (Photograph by Whitney Barlow Robles; by permission of the Trustees of the Natural History Museum, London)

fish command a watery world that is not ours, one that covers more than 70 percent of the planet, even as an eel, like a diving human, can moonlight in a hostile element by snaking across land in a pinch.[36]

Many ocean creatures presented challenges for the fish herbarium unique to their species. The most interesting feature of the "sucking fish," or remora, was (still is) the flattened suction pad atop its head that lets the fish fasten to larger bodies, be those shark bellies or ship bottoms (figure 5.10). This curious and highly modified organ, an evolved dorsal fin, was of vital interest to natural history for the fish's reputation as quite a drag. In antiquity, such animals were widely thought capable of latching onto vessels and slowing or stopping their movements, as seen in Pliny the Elder's account of remoras. They might so decide the fates of battles and empires; the word "remora," from Latin, means "hindrance" or "delay." Although many eighteenth-century naturalists such as Mark Catesby would write off the fish's ship-stilling powers as "fabulous," remoras took on renewed significance in the context of New World nat-

Figure 5.10. Remora from Marcus Éliéser Bloch, *Ichtyologie ou Histoire Naturelle, Générale et Particulière des Poissons,* Part 5 (1787). (Biodiversity Heritage Library)

ural history. Naturalists sought to understand remoras' relationship and possible cooperation with sharks—and the mystery of why sharks declined to eat these fish. They also reported that Indigenous Caribbean fishers would collaborate with remoras to secure turtles and manatees. But as for fish flattening, the animal's storied organ posed a conundrum: it was horizontally flattened in its natural state in contrast to the animals' vertically compressed bodies. This anatomical problem forced naturalists to weigh trade-offs between different modes of presentation. Garden decided to send Linnaeus whole, dried specimens of the animals rather than slitting them in half like his other flat specimens; in contrast, a remora in the Gronovius collection has been halved and mashed to fit on the page, the suctioning organ unnaturally turned upward.[37]

Other creatures, round and radiating when alive, seemed wholly unfit for paper, as with the aptly named lumpfish in Peck's collection. In some cases, the fish's dramatic proportions might have aided its journey to becoming a specimen. The Salem clergyman and collector William Bentley, a correspondent of Peck's, was almost cheating when he collected a lumpfish that unluckily found itself stuck and exposed between two rocks right as the tide receded. But flattening was hardly the obvious choice for preserving a bulging lumpfish in perpetuity.[38]

The naturally compressed form of Peck's flounder, on the contrary, nearly beckoned preservation on paper. And yet, even a flatfish like a flounder or turbot or plaice presented challenges for two-dimensional preservation. Extant specimens retain pools of varnish on their skins where fingers once pressed, and a fine cutting maneuver with little margin for error would have been required to slice the animals apart. Moreover, George Shaw of the British Museum noted the unique difficulties of printing images of flounders and their relatives in books. Shaw claimed that "nature exhibits a most extraordinary deviation from her usual plan" in these animals by breaking from the typical body format of symmetry. In fact, the placement of a flounder's eyes—either on the right or left, from a human's top-down perspective—denotes its species designation in the Linnaean system. This peculiarity forced printmakers to incise copper plates with the specimens drawn *backward* in preparation for printing, with taxonomic consequences should this step be neglected. Shaw thus warned: "If the engraver is not careful to reverse the drawing, it will give the species in a wrong division of the genus."[39]

Another group of flattened fish—the rays—frustrated the very visualization machinery of the fish herbarium, with its simple and symmetric substitutions. They stood as winged warnings of what went missing when people tried to crowd nature into a book. A stingray specimen cannot be shown in profile. It would appear as little more than a line. In the case of most fish, one side of the animal serves as a proxy for the other when flattened on paper. But all representations of rays devoted to capturing the whole creature and the diagnostic features on both surfaces must be double images (figure 5.11). Peck left his own dual drawing of a ray among his lecture notes, with the top of the animal shown on one side of the paper and its belly on the other; this depiction required that he flip the drawing like a coin, or a page, to showcase the full creature to students. Through their adaptations, life practices, and physical configuration, rays subtly subverted the efficiency of the bisected fish herbarium specimen.[40]

That doesn't mean rays weren't intensely studied in the eighteenth century. Fish preserved on paper were visual technologies. Rays and their close relatives, in contrast, opened up whole alternative worlds of tactile engagement and embodied interaction with animals. In particular, craftspeople created a sumptuous substance called shagreen by processing ray, skate, and shark skin into a striking beaded leather: they filed down the animals' tooth-like scales, called dermal denticles or placoid scales, and dyed the resulting mate-

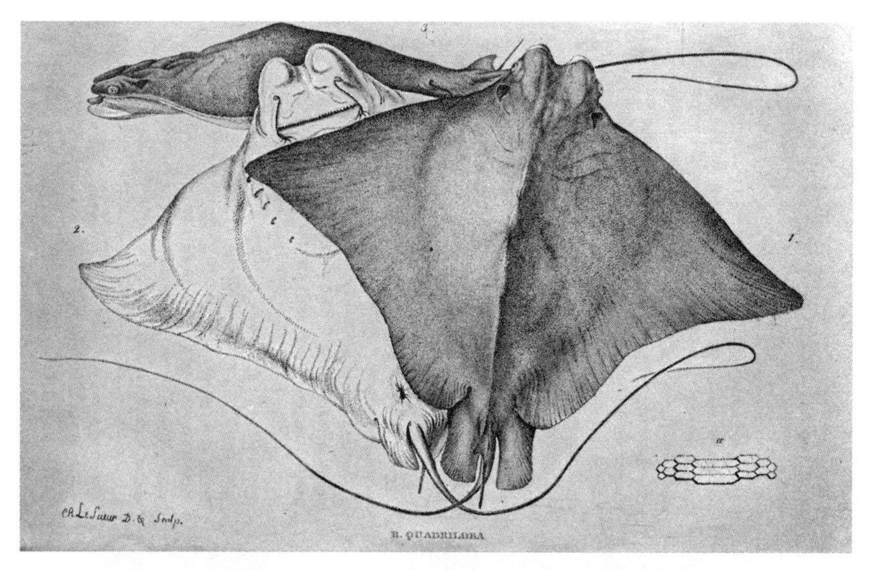

Figure 5.11. Three views of a cownose ray (*Rhinoptera bonasus*). (From Charles Alexandre LeSueur, "Description of Three New Species of the Genus Raja," *Journal of the Academy of Natural Sciences of Philadelphia* 1 [1817–1818]; Biodiversity Heritage Library)

rial a rich gray, red, green, or, least commonly, blue (figure 5.12 and plate 10). Shagreen was flatness refigured.

Descriptions of shagreen artifacts tended to celebrate the human skill behind them. For instance, Oliver Goldsmith's popular treatise noted that the fish skin that served as shagreen's basis was "by great labour, polished into that substance." Scientific instrument makers used this coarse but visually appealing animal matter to wrap scientific instrument cases. It also covered the barrels of optical apparatus like microscopes and telescopes, offering a sturdy and durable grip for adjustments of focus. Shagreen served a similar purpose in Japan by covering the hilts of samurai swords; it has also traditionally been used there to grate wasabi.[41]

Similarly, mapmakers took rays' once-flat skins and rolled them over the cases of pocket globes, lest the orb slip from a philosopher's hand (figure 5.13, figure 5.14, and plate 11). Although rays evaded the convergence of animal and image allowed by the fish herbarium, they entered the practice of natural history and natural philosophy in profoundly tangible ways. In so doing, they suggested the limits of purely visual or rational approaches that treated wildlife in

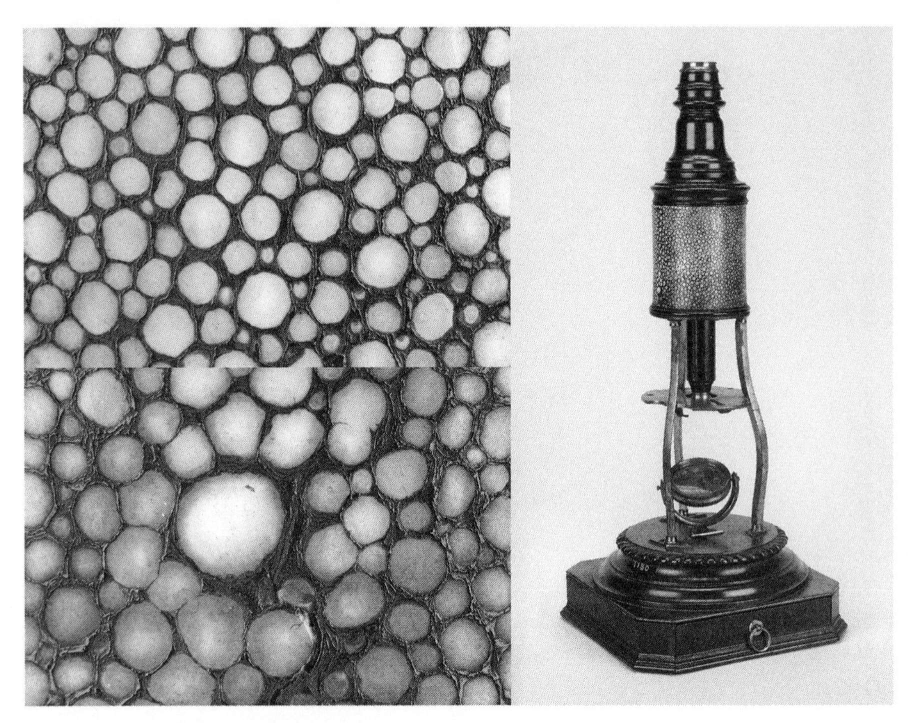

Figure 5.12. Matthew Loft, Culpeper-type compound microscope with rayskin shagreen covering (and magnified details), 1730–1747, the Collection of Historical Scientific Instruments, Harvard University, 1180. (Courtesy of the Collection of Historical Scientific Instruments)

flattened, book-like terms. If Peck's delicate dried fish whispered *don't touch,* a stingray converted into a handle demanded contact, suggesting the need to grasp material nature rather than simply apprehending it optically or cognitively. As a way of *feeling,* these animals facilitated ways of *seeing* the world, from the whole earth in miniature, to the moons of Jupiter through a telescope, to the smallest flea pressed flat on a microscope slide.

Microscopes, moreover, offered something paper fish did not: transparency. They gratified the fantasy of seeing the whole specimen at once, letting naturalists peer beyond surfaces, into interiors normally hidden from sight, circumventing illustration's dilemma of one-sidedness. Already in 1665, Robert Hooke had extolled the wonders of "internal curiosities" in his breathtaking *Micrographia,* a treatise on invisible worlds lain before human eyes by the microscope. Peck certainly found God on smaller scales, calling insects that he examined "puny vouchers of omnipotence" after the poet Edward Young.

Figure 5.13. John Cary, pocket globe with sharkskin shagreen case, 1791. (National Library of Australia, 5213930; photograph by William Robles)

In the first half of the eighteenth century, the Londoner Matthew Loft was one among many purveyors to produce handsome shagreen-wrapped microscopes in the style of instrument maker Edmund Culpeper. Included among such a microscope's accessories was a metal cradleboard contraption called a fish plate or frog plate, one of which bears traces of extensive use (figure 5.15). Three straps of fabric would confine a live fish or frog to the metal plate so that one could examine, with the microscope, blood coursing through the overhanging tail or foot. The discovery of the circulation of blood in the early seventeenth century had created a demand for instruments that could capture the movement of internal life processes in real time. As George Adams wrote in his *Micrographia Illustrata,* "The circulation of blood affords an entertaining sight," and, "the smallest creatures are perfect in their kind, and carry about them as strong marks of infinite wisdom and power, as the greatest." Curiosity relished object lessons at lesser scales. Eighteenth-century naturalists restrained nature

Figure 5.14. John Cary, pocket globe with sharkskin shagreen case, 1791. This particular pocket globe features tracks of Captain Cook's voyages. (National Library of Australia, 5213930; photograph by William Robles)

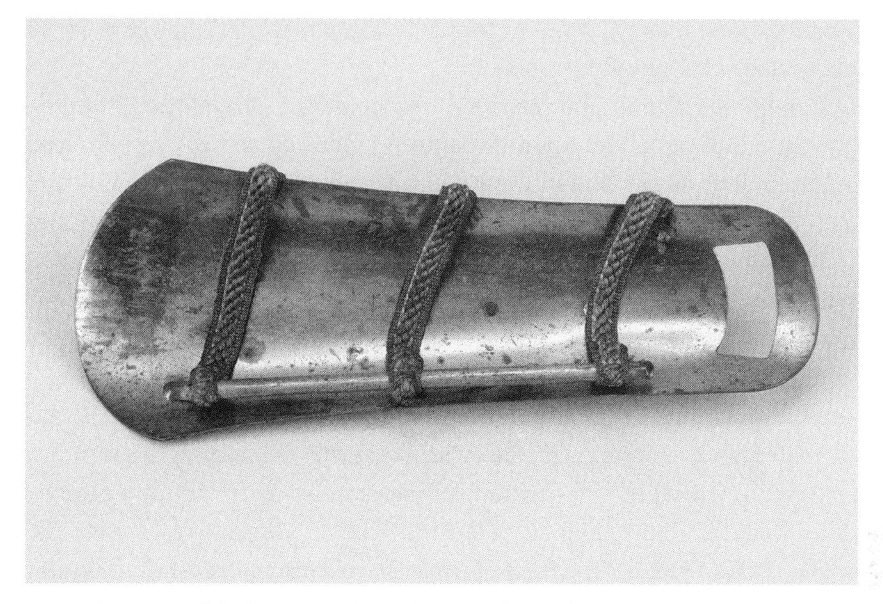

Figure 5.15. Frog or fish plate accompaniment to Matthew Loft's Culpeper-type compound microscope, 1730–1747, the Collection of Historical Scientific Instruments, Harvard University, 1180. (Courtesy of the Collection of Historical Scientific Instruments)

through compression on slides and fish plates, grasping ray skin in their hands to see more holistically the covert operations of animal life.[42]

Indeed, rays and their relatives entered scientific practice and European and American collections in a variety of visceral media. Sensation and texture were of central importance when studying the creatures. Collections in Harvard Hall featured drawings of rays alongside shagreen-wrapped instruments, and the early museum of the Massachusetts Historical Society featured a number of objects from the Pacific world, including "the Bone of a Sting-ray, formed into the point of a war spear, from the Pelew Islands." Joseph Banks, while touring the Pacific on Captain Cook's first voyage, claimed that "nothing is more terrible" to Europeans than the menace of a stingray's barbed tail. He meanwhile found the mouthfeel of stingray meat "abominably coarse." The very term "shagreen" comes from its roughness, being a variant of the word "chagrin," defined by the Oxford English Dictionary as signifying, in the eighteenth century, "acute vexation, annoyance, or mortification, arising from disappointment, thwarting, or failure." The English word derived from the French *chagrin,* whose material sense of "rough and granular skin" evolved to metaphorically evoke "gnawing trouble."[43]

The tactile (and at times negative) connotations of shagreen could figure into appraisals of the character of shagreen-producing animals, as intimated by Banks. Rays, skates, and sharks are all what biologists call cartilaginous fish, having an internal armature made of cartilage instead of bone. Certain species of all three could surrender shagreen, though that wrought from sharkskin bears a slightly different, more uniform appearance. The bristling associations shagreen caused in people, deriving from touch, surfaced in period depictions of rays and sharks as living beings. Goldsmith, in addition to lamenting a shark's "great goggle eyes, that he turns with ease on every side," noted in horror that "his whole aspect is marked with a character of malignity: his skin also is rough, hard and prickly; being that substance which covers instrument cases, called shagreen." Rays often didn't fare much better in such assessments. The angler William Hughes instructed people to avoid eating many rays, first casting aspersions on the eagle ray as having "a most repulsive appearance" and then calling the "sea devil," likely referring to a devil ray or manta ray, "a perfect marine monster, owing a still more fiend-like tail, whilst its expanded wings and cloven snout seem to give it a fair claim to the diabolical title that has been conferred upon it." Bodies of loathed and feared sharks and rays became domesticated and then instrumentalized in the naturalist's cabinet. Yet the perceived rough and dangerous character of the animals seemed to live on through the materiality of their skin and bones.[44]

Fish that yielded shagreen were intelligent, capable, and formidable, ready to strike humans in their way (figure 5.16 and plate 12). When alive, one group—the electric rays, especially those of the genus *Torpedo*—became the object of intense, haptic study. These fish deliver potent shocks to stun prey or predators. Their torpifying effects had for centuries earned them an association with natural magic and occult powers. Such animal electricity fascinated philosophers who were also investigating man-made forms of electricity in the eighteenth century using apparatus like spinning electrical machines, glass Leyden jars to store electrical charges, and eventually model torpedo rays made of leather and pewter to simulate the animal itself. Without these creatures, we might not have the electrified world in the form we know it today.[45]

The British experimenter John Walsh closely studied and felt the shocks of torpedo rays during a research sojourn in France, whose surrounding waters teemed with the animals. In the early 1770s, he penned a series of letters

Figure 5.16. Eagle ray (*Aetobatus ocellatus*) in watercolor and pencil by Johann Georg Adam Forster, dated May 10, 1774, Tahiti. This image was created during Captain Cook's second voyage (1772–1775); the Māori word for ray, *whai,* appears in the left margin. This ray, in Forster's hand, takes on a sense of movement and near-personhood in line with period conceptions of rays and sharks as particularly formidable and willful ocean denizens. The unfinished illustration—to which only a few of the eagle ray's characteristic spots have been added—opens a window into Forster's artistic process, as seen by the instructions "Dark Red colour" written under the animal's left pelvic fin. (By permission of the Trustees of the Natural History Museum, London)

describing his discoveries to his fellow electrophile Benjamin Franklin, which the *Philosophical Transactions* published in modified form. Walsh employed his body as an instrument to fathom how such animals might "throw light on artificial electricity." Indeed, it was Walsh who definitively demonstrated that the pain and numbing sensations caused by the fish were electrical rather than mechanical in nature. To Walsh, the creature was no passive scientific object. He marveled at the "will of the animal" and its "state of liberty" as it exercised its "power," visibly "winking" its eyes as it delivered shocks when disturbed. And disturbed they were by his probing: part of Walsh's research program involved grabbing an electric ray and violently dunking it in and out of water for a full minute. Walsh even formed séance-style human chains with other company, letting the fish's electrical powers flow through all those present.[46]

The two-sided nature of rays that had frustrated their conversion into paper specimens here proved vital for understanding the animal, as well as the polarity of electricity. Walsh stressed the importance of studying the entire ray, noting that "we have discovered the back and breast of the animal to be in different states of electricity" and that "their upper and under surfaces are capable, from a state of equilibrium with respect to electricity, of being instantly thrown, by a meer energy, into an opposition of a *plus* and *minus* state, like that of the charged Phial [Leyden jar]." Moreover, their internal electrical organs, composed of hexagonal columns that stretch from the top to the bottom of the animal in kidney-shaped clusters, were essentially vertical natural batteries (figure 5.17). In fact, they paved the way for the invention of batteries. At the end of the eighteenth century, Alessandro Volta explicitly modeled his Voltaic pile—the first battery to provide a continuous electric current—on the electrical organs of torpedoes and other electric fish such as New World eels. In his account of the ray, Walsh flipped the oft-repeated metaphorical parallel between man the watchmaker and God the watchmaker by equating human and animal artisanship, writing: "He, who analysed the electrified Phial, will hear with pleasure that its laws prevail in animate Phials: He, who by Reason became an electrician, will hear with reverence of an instinctive electrician, gifted in his birth with a wonderful apparatus, and with the skill to use it."[47]

Yet Walsh also acknowledged barriers to understanding the full mechanics of the torpedo's electrical apparatus, concluding: "We here approach to that veil of nature, which man cannot raise." More practically, the shocks of torpedo rays were also said to be stupefying. They temporarily left naturalists mute and unthinking. William Bullock reported that the animal's power caused "for

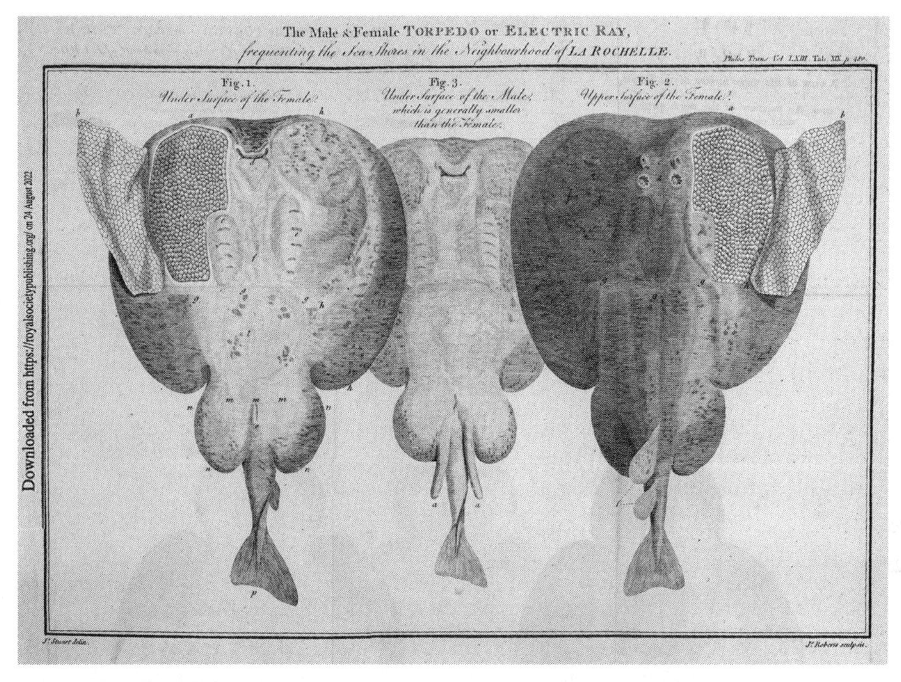

Figure 5.17. Male and female torpedo, showing both upper and lower sides of the fish and their columnar electrical organs that stretch from the bottom to the top of each animal in large clusters. (Printed in John Walsh, "Of the Electric Property of the Torpedo," *Philosophical Transactions* 63 [1773]; Royal Society of London)

a while a total suspension of the mental faculties." Such rays could not simply be observed. They had to be felt, even at the risk of briefly losing one's own mind.[48]

Naturalists couldn't help but recognize these animals' power—both in a literal electric sense and also in the sense of the creature's efficacy and willful warning. Walsh and others conceded the hubris of any mission to lay claim over the animal kingdom, leaving nature's veil partly closed. Rays and their sensational bodily effects were not a fringe interest in the study of fish. They were some of early ichthyology's central curiosities. George Humphrey beseeched his collectors to paste numbers directly onto fish specimens that would correspond with memoranda in a notebook, where he likewise urged them to record for each specimen "any curious particulars concerning it that may occur to you. For instance 'The Torpedo, or numb fish when touch'd with the hand benumbs all that side of the body of the person who touch'd it.' 'The Sting Ray, when attack'd endeavours to strike his Enemy with his Sting, which

Figure 5.18. Eighteenth-century torpedo ray (*Narke capensis*) specimen from the Gronovius collection at the Natural History Museum, London. (Photograph by Whitney Barlow Robles; by permission of the Trustees of the Natural History Museum, London)

being set with sharp beards on each side the whole length thereof, makes the extraction of it worse pain than the insertion, and makes the wound dangerous.' and so on." As Humphrey's examples make clear, the study of rays was multisensory, tactile, and rooted in bodily sensation. It was rooted in the reality that naturalists were animals too. Rays showed one could not simply read the book of nature: one also had to, quite literally, get a grip on it.[49]

A flattened specimen of a torpedo ray resides today in the Gronovius fish collection in London, its "winking" eyes still prominent, though the rest of its body is flattened like a pumpernickel pancake (figure 5.18). Despite how unsuited rays were for the fish herbarium, a handful of imperfect specimens remain. The essential features of this ray's underbelly were severed and lost to history in favor of converting the animal into another page of nature, easy to file in a box. Now dead, it has been neutered of its famed electrical powers. Touched with a gloved hand, it delivers no jolt. Even so, the ray provides a record and reminder of a specific nonhuman individual who came under an eighteenth-century collector's gaze (and knife). Its skin survived, as one flash of that history, to tell one side of a two-sided story.

Specimens. Skins. Paper. Microscopes. Maps. Electricity. Fish. What causes these to hang together? Recall from our embarkment the 1775 portrait of Laurens Gronovius, son of the same Gronovius who published the Atlantic world's most popular fish flattening formula. In that painting, Laurens writes at a table while surrounded by books and manuscripts, a Culpeper-type microscope wrapped in rayskin shagreen, and various apparatus for electrical demonstrations. Fish, corals, a two-legged salamander known then and now as a siren, and other curiosities preserved in dry and wet forms line the mantel and spill out of drawers and cabinets. Here, natural history is a family affair: Laurens, who inherited and augmented a collection from his naturalist father, now transmits such knowledge to his children through their tactile engagement with specimens and books. With a quill in one hand, Laurens presumably describes the dried and flattened butterflyfish he clutches with the other. Meanwhile, his eldest displays a botanical illustration—a literal page running parallel with the compressed fish specimen, a figurative page from nature's book. The round globe, turned to the New World, looms over this act of transmitting flat fish into flat text, nodding to the imperial reach of natural history collecting. Such an image connects the dots of eighteenth-century science by gathering its diverse activities in one unified space. The painting is powerful visual proof of how the study of flat and round animals, plants, geography, electricity, fish skin, and the microscopic world were joined together under a single scientific practice, despite how disjointed these endeavors might seem to twenty-first-century eyes.

This two-dimensional image, full of two-dimensional practices, erases the vast landscape of human actors and expertise that built the Gronovius collection. Its animals likewise appear passive, pliant, and dumb. By the time a new cohort of American writers emerged in the first half of the nineteenth century, rejecting the visualization and rationalization of Enlightenment-era science in favor of wide-eyed awe at nature's sublimity, something else about Gronovius's cabinet, this flattened philosophy of seeing animals, would seem decidedly off. In his novel *Moby-Dick* from 1851, Herman Melville parodied the metaphor of the book of nature. The infamous "Cetology" chapter absurdly classifies whales in the lexicon of book printing, referring to the animals as folio, octavo, or duodecimo, according to their size—the implication being that one does not come to truly know whales *from* flattened books or *as* flattened books, but only by entering the ocean to encounter them.[50]

Two centuries prior, Hooke's larger-than-life foldout plates of fleas and lice leaping from his magnificent *Micrographia* unknowingly presaged Melville's critique of a textual natural world (figure 5.19). They made the transmutations of the flattened microscope slide visible, rendering giant the small, the unseen, the discreet. A volume proved too modest to contain God's text, bearing

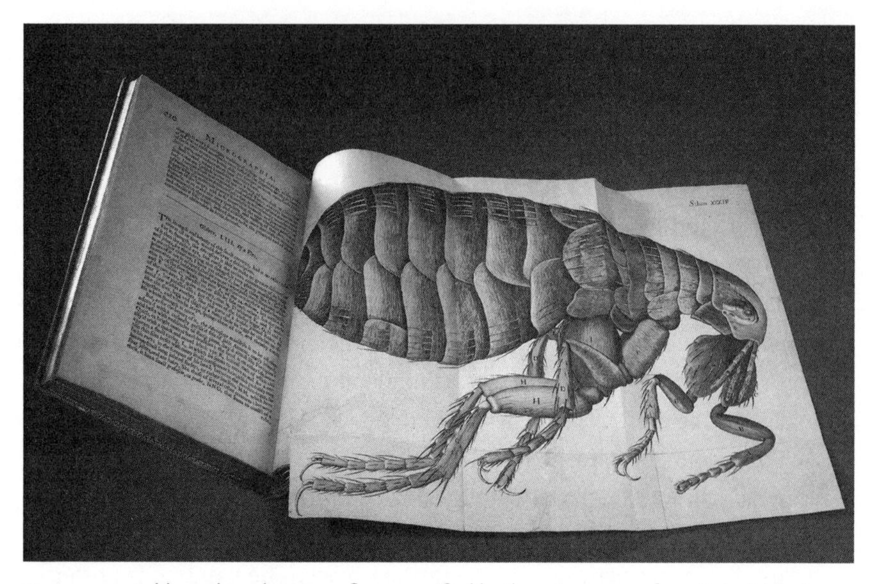

Figure 5.19. Foldout plate showing a flea magnified by the microscope, from Robert Hooke, *Micrographia* (London, 1665). (Wikimedia Commons)

monstrous magnifications that exceeded the metaphor of nature's book and questioned the ability of flattening, whether through images or through specimens squished onto slides, to make that metaphor concrete. Oliver Goldsmith tempered his peers' celebration of the microscope by warning in the 1770s: "Too close a pursuit of Nature chiefly leads to uncertainty." Little more than a half-century later, an exasperated Henry David Thoreau called museums and cabinets like that of Gronovius the "catacombs of nature." He wrote in his journal: "Where is the proper herbarium, the true cabinet of shells, and museum of skeletons, but in the meadow where the flower bloomed, by the seaside where the tide cast up the fish. . . . Would you have a dried specimen of a world, or a pickled one?"[51]

Scholars such as Susan Scott Parrish have uncovered Indigenous, Black, female, and lay affronts to natural history's ordering and reasoning much earlier than the writings of Melville and Thoreau. But there was, too, a very different sort of challenge to natural historical methods—what we might call an animal critique, and one that eighteenth-century naturalists knew all too well. Naturalists tried, ham-handedly, to cram nature into a book. But fish and other creatures circumscribed what could be visualized, stabilized, and known. For all their desires to gather up the world, naturalists also admitted that the earth, its animals, plants, terrains, and curiosities remained beyond their full comprehension. Natural history could be a deeply embodied and sacred act that offered both a wondrous mutability and a form of immortality. With baptismal resonances, William Swainson made sure each flattened fish he prepared was "well anointed with the preservative"—born again as a specimen. Indeed, Thoreau's remonstrance missed the transformative power of animals to reach past the dryness and pickling: to live on, even in death.[52]

CHAPTER SIX

Splitting the Lark

THERE IS BLOOD ON MY HANDS.

Catching and killing animals, gathering their corpses, transmuting them into specimens, and maintaining their bodies in catacombs for years after death—all the many material motions of natural history—required intimate bodily interactions with nonhuman flesh at every turn. These doting human bodies possessed knowledge. Much of that knowledge, however, was ineffable. In the words of the theorist Michael Polanyi, emphasis his, "*we can know more than we can tell.*" In recent years, historians of science like Pamela H. Smith and H. Otto Sibum have instrumentalized their own hands to explore how certain physical practices can't be conveyed in words. Historical fingers and torsos and eyes possessed "tacit" or "gestural" knowledge that seldom translates well into linguistic form. Much of it is not even fully conscious. This experience and tacit knowledge were fundamentally different types of knowing nature than written ones, as Smith has argued, requiring distinct methods of investigation and analysis. In other words, one way to understand historical making is by doing.[1]

All that a fancy justification for taking a life.

For some time, I had resolved to try my own hand at following Gronovius's fish flattening instructions as published in the *Philosophical Transactions* in 1742, given the ubiquity of that recipe and my hunch that William Dandridge Peck had followed it as well. But I was still summoning the gumption to do the thing. Having read every source I could find on the process, I wanted to know more: was it really as simple as instructions promised? What steps and gestures weren't and perhaps couldn't be conveyed in words? What know-how did I lack? What context, as an amateur-bordering-on-dilettante of the twenty-first century, did I lack? And would I really be baffled by a dead fish, powerless to preserve its semblance for nature's book? Zach Rotholz, a bosom friend from college and mad-scientist engineer, was visiting one August weekend, and he finally convinced me that flattening a fish wouldn't just be edifying: it might also be fun.

I didn't kill the fish, though its literal blood would be on my literal hands soon enough. But I might as well have. As an on-and-off-again vegetarian for much of my adult life, I've killed many fish by proxy simply by eating them. Sometimes, I've tried to stop the carnage. While navigating the Great Barrier Reef with John and Shane, we stilled the boat for some fishing, according to strict sustainability practices. Using a handline, I heaved up, through great struggle, a hefty coral trout. The Australians were ecstatic at its size. But as the fish hung vertically on the line and I stared down into the black hole of its mouth, into its very core, yellow vomit spewed dramatically from its belly. Was it hurt? Scared? Simply suffocating in air? I stood stunned, as both men promptly took and dispatched the fish. My expression must have changed. "Oh—you *did* want to eat it?" they asked. I gave half of the catch to them, and William and I cooked the rest back in our cabin that evening. The entire supply chain was as local and humane as turning a life into food could be. But as I swallowed the fish's wonderfully buttery flesh, and a kookaburra laughed in the trees, all I could taste was that dying animal's bile.

It's easier not to think about death when you're at the grocery store. You can choose among already lifeless animals or barely recognizable animal parts—which, to be fair, would also have been customary for naturalists who procured ample specimens from fish markets in the eighteenth century. With Zach's emboldening, we arrived at the local Whole Foods. I asked an incredulous fishmonger which specimen might be simplest to halve and flatten on paper like they did in the olden days. He warily recommended a whole European sea bass, quite manageably sized. *Dicentrarchus labrax* to scientists and *branzino* to many a restaurant-goer, this plucky predator and sometimes-cannibal was small enough to fit on a sheet of butcher paper, which we acquired over at the cheese counter. These days, the mild and sweet sea bass is trendy ocean fare, farmed from land-based tanks or sea cages plopped directly into the

Mediterranean and elsewhere throughout Europe's waters. It became one of the first fish species to be grown commercially there given its littoral love of lagoons and estuaries and potential for mass production of juveniles. The fish's ubiquity and popularity—despite conservation concerns and declining wild numbers—made it particularly accessible to us this day for specimen-making and knowledge-making.[2]

My heuristic hack job revealed unwritten dimensions of labor behind flattening and specimen creation. It illuminated the active afterlife of the sea bass and how animal bodies guide what human hands and eyes can do. Most of all, it generated new unknowns and further questions. It was rife with mishaps, resulting in a specimen that is, in certain ways, laughably unlike the eighteenth-century examples I study. Then again, everything seemingly wrong about my fish also underscores the improvisational nature of natural history in early America and across the Atlantic world. There was no one way, or right way, to flatten a fish. In colonial settings or budding republics, where many of these animals were first collected and fixed to paper, people had to make use of the materials available to them. They flattened fish on the fly—which might mean borrowing prep materials from the kitchen or sometimes making a fish into a much-needed meal instead of an artifact in a far-off naturalist's cabinet. What was more, I began to suspect that naturalists' breezy assurances about the ease of the flattening process actually erased the craft knowledge, as well as the often unfree labor, that produced these animal artifacts.[3]

With a printout of Gronovius in hand on that hot summer day, I brought our newly acquired specimen-to-be to my apartment complex's communal kitchen. Zach made for the ideal sous-chef, offering encouragement and strategic suggestions as I took the lead. Somehow, I knew that an engineer who had founded his own cardboard furniture business and who, in our college days, figured he'd float downriver in a raft made of soda bottles and other trash and rigged countless other mechanical contraptions in our dorm—like turning clothing hangers and twine into a needlessly elaborate pulley system that stretched the length of the room and allowed one to switch off the lights without rising from bed—would be a fitting partner in crafting crime.

First, I followed Gronovius's command to hold the fish with my left hand, its head pointed at my belly. I bisected the animal's body with "Scissars," being careful to keep all midline fins on one half of the fish in anticipation of the

Figure 6.1. Splitting the sea bass specimen. (Photograph by Zachary Rotholz)

final product. However, I met with resistance at the head. Here already I deviated from the script and split the skull in two with a chef's knife (figure 6.1). Perhaps my scissors lacked the "very fine Blades, and sharp Points" urged by Gronovius. The next near-hour involved tediously removing bones, gills, organs, wet membrane, and flesh—so much flesh—despite Gronovius's claim that much of this would be accomplished "easily." Holding the whole fish in my hands and then navigating its internal anatomy was just as compelling and intimate a reminder of animals as historical beings as touching finished specimens with gloved hands in an archive has been. Once I felt that I had sufficiently separated the sea bass's skin from everything it had contained in life, I spread what was left of the animal on a board and expanded its fins with pins, as instructed by Gronovius, to keep those crucial Linnaean characters countable (figure 6.2). Trying not to alarm neighbors and lacking sunlit outdoor space of my own or a warm open fire (the recommendations of my guide), I opted to dry the still-moist fish skin on low in the oven, knowing early American

Figure 6.2. The still-wet specimen, its fins expanded with pins in preparation for the drying process. (Photograph by Whitney Barlow Robles)

naturalists like Manasseh Cutler, among others such as René-Antoine Ferchault de Réaumur and Sarah Bowdich Lee, suggested preserving animals, sometimes between sheets of paper, in an oven "after the bread is drawn."[4]

Once the specimen was dry, Gronovius said that it "must be pressed flat." At this juncture, I soon realized the difficulty of keeping the fish so, and, by extension, the conscious investment placed in flattening by eighteenth-century naturalists. Many historical specimens I have observed are thinner than cardstock and astoundingly, impressively flat. Some are flush with the paper, with barely a bubble, ripple, or ridge. My specimen, on the contrary, continues to curve back to its living form, while its baguette-like crunchiness, also evident in a large crack, prevents me from fully flattening it after the fact (figure 6.3). Practicing over time, with additional specimens—even better, with a flounder, lumpfish, and ray—would yield more opportunities to conquer this step of the process through experience. But for my part, that would also require taking additional lives by proxy.[5]

My fish's unflatness did more than simply suggest flaws in my own tech-

Figure 6.3. Dried specimen of *Dicentrarchus labrax* (European sea bass). Animal specimen and ink on paper. (Photograph by Whitney Barlow Robles)

nique, however. It also helped me intuit the extended presence of the naturalist, or some shadowy helper figure, after the initial cutting steps. I likewise gained a better grasp of the fish's prolonged presence as humans worked to bend these vital materials to their will. Some versions of instructions do stress the customizable nature of fish herbaria. The English physician John Coakley Lettsom, who supplied Harvard's early collections with various natural history specimens, deviated from the strict flatness paradigm in writing that fish skins could be "dried upon paper like a plant, or one of the sides may be filled with plaster of Paris, to give the subject a due plumpness." Other instructions also propose stuffing the skins with tow. Despite room for variation in form, however, the deliberate flatness I have witnessed in so many specimens—the rule rather than the exception—seems all the more remarkable and coerced now that I have tried my own hand at the process and wrestled with a fish body hell-bent on curling back to three dimensions, to the contours of its former life.[6]

Caretakers must have constantly attended to these stubborn materials through vigilance and persistent pressing—something I, quite frankly, didn't budget time to do amid other commitments, including my base desire for sleep. It is hard to appreciate how protracted this procedure was from a page and a half of sparse instructions, though Gronovius did intimate the incessant management of leaky specimens when he cautioned: "as a sort of glutinous Matter, in pressing, is always forced out from betwixt the Scales and the Skin, a Piece of Parchment is to be laid under the Fish, which is easily separated from the scales, but Paper always sticks: For this Reason it is necessary, that after an Hour or two, a fresh Piece of Parchment should be applied." This, he noted, would prepare the fish "in the Space of 24 Hours," implying frequent parchment swaps throughout a full day and night of observance. If one simply entombed the skin in a heavy volume or other mechanical flattening arrangement such as a botanical press, the oozing fish might irrevocably cling to its surroundings in the absence of frequent intervention. In Gronovius's experience, and now mine, the fish continued to assert its sticky (and stinky) existence.[7]

By preserving my own fish, I also began to understand the risk, and perhaps the apprehension, that attended specimen care and creation: scissors and knives and pins might slip and injure the specimen or the preserver's hands, and fish bones can be as sharp as scissors, as eighteenth-century cookbooks often warned. I escaped with only one small injury. This is also a form of destructive sampling, and an irreversible act: naturalists had to carve up what was often the only specimen of that species they would see in their lifetimes—and, in some cases, an animal entirely new to European science. They had one chance to bisect the specimen correctly.

Making and then monitoring a specimen also helped me understand how it changed over time, alerting me to issues I did not know existed. For instance, although naturalists constantly bemoaned the loss of color in their fish specimens, mine has retained its blue hue for several years, and it only recently started to yellow from the bottom up, revealing a much more protracted process of transformation than writings about these specimens often capture. Perhaps this derives from differences in preservatives used, none being mentioned by Gronovius in his published tract. I settled on a probably toxic varnish from the hardware store, rather than a probably toxic eighteenth-century recipe, though I subsequently found instructions Peck consulted and supplied to the Massachusetts Historical Society, which read: "Take twelve ounces of *rectified spirit of wine;* one ounce and a half of *spirit of turpentine;* mix, and add half an ounce of *camphor.* The skins of animals may be passed over with this fluid, by means of a brush. It will destroy several species of insects." Turpentine would destroy me, too, were I to imbibe or deeply inhale such a mixture.[8]

Living with the skin of a fish over time has, moreover, revealed the anxiety of maintenance. I kept my specimen atop my tallest bookshelf and away from a curious cat for months, but, as a result, when I went to examine it one day I dropped it from six feet in the air and lost a pelvic fin in the process. For easier display, retrieval, and long-term storage, and lacking any sort of formal archival system in my home, I ultimately opted to keep it in a shadow box designed for military medals rather than on paper—a decision which in itself reveals how paper fish herbaria did not always enable simple displays and functioned more as reference and teaching specimens. Hanging in a wooden frame on my wall, where it has been, at the time of writing this, several years and counting (yes, books take a long time), my specimen doesn't look too far off from an animatronic singing bass. Peck noted in a lecture that he would "lay before" his students some fish specimens, suggesting that, much like a book, they did not lend themselves to vertical exhibition while stored on paper. Instead, they encouraged top-down imperial observation over a table.[9]

Standing in a kitchen with a knife in one hand and a fish in the other, the act of making also prodded me to begin asking a new set of research questions about my scientific material, ones that were now culinary and cultural in nature. Would a historical chef fillet a fish the same way I do, I wondered? Would she eat it with the skin on? Did food preservation resemble specimen preservation? Was cooking the model, implicitly or explicitly, for creating these specimens? What were the gender politics of this food-made-flatness, given that natural science was largely (though not entirely) closed to women at this time, yet women, as well as male servants and enslaved people of all genders, did much of the cooking and would have intimate knowledge of fish interiors? Understanding how fish were prepared for the table and fried up as meals might also indicate what unstated precedents or tacit knowledge the creators of fish herbaria drew upon as they slit animals in half and placed them on paper—paper that must have, to some, resembled a plate.

These questions sent me back to the books—this time, historical cookbooks, opening my eyes to how angling and cooking might themselves be considered modes of natural historical practice. Fishing required deep knowledge of species-specific fish habitats, behavior, and development. Preparing fish as food involved collecting, identification, classification, and implementation of best practices for drying and preservation. All were likewise tools in the naturalist's repertoire. Some cookbooks, like *Fish, How to Choose and How to Dress,* published in 1843 by William Hughes (alias "Piscator"), not only drew on decades of fishing knowledge but also explicitly cited naturalists and ichthyologists such as William Yarrell and Edward Donovan.[10]

Hughes gestured toward the hands-on knowledge behind prepping and cooking fish as he promised to "treat of cleaning and preparing fish for cookery, a very important affair, though one to which very little attention is usually paid." Hughes explained that absence by noting the work was "generally delegated to unskilful hands, who take little pains or trouble to execute the task in a masterly manner, the consequence of which is, that many excellent fish are all but spoiled." These subaltern fish artisans no doubt harbored more expertise (tacit or otherwise) than Hughes conceded. His dismissal may have been, as much as anything, a rhetorical move to boost his own authority and present the subject matter as uncharted territory. In the book's second edition, titled *A Practical Treatise on the Choice and Cookery of Fish,* Hughes further clarified that these supposedly "unskilled" hands often performed the task of fish preparation because it was "a disagreeable office." In other words: cutting a fish open is gross. It required a date with organs, bile, and blood. Consequently, fish gourmands historically relegated such a task to laborers who were lower on the social hierarchy, removing themselves from the embodied act. The carnage frequently fell to wives, servants, and enslaved cooks, stretching from Hughes's day back into the eighteenth century and earlier.[11]

The second edition of Hughes's book also included instructions that resemble Gronovius's process, showing implicit parallels between kitchen craft and natural historical practice. "Fishes that require to be opened by the back," wrote Hughes, "should be split through from nose to tail with a very sharp knife, so that the flesh may be cut clean and without being jagged close to the back bone." Numerous women throughout the Atlantic world offered their ichthyological expertise by publishing widely distributed cookbooks in the late eighteenth and early nineteenth centuries as well, including Maria Eliza Rundell. In *A New System of Domestic Cookery,* first printed in Britain in 1806 and in America in 1807, Rundell offered similar Gronovius-style advice, including: "After scaling and cleaning, split the salmon." And she suggested tailored methods of dry preservation for different species of fish.[12]

The shared manual maneuvers present in the kitchen and curiosity cabinet required analogous mastery of fish bodies, even if that anatomical know-how would ultimately serve different ends. With this new alimentary awareness—the by-product of gutting a fish in a kitchen—I returned to some of my more strictly scientific sources. Glimpses of food and culinary techniques began to stand out like pan fires. They had been present but largely invisible to me before. But indeed, scientific texts about animals often took on traits of cookbooks. For example, it was customary for naturalists to note the larger cultural meanings of animals in their accounts, including whether certain species made for good food. Cookery and natural history frequently borrowed materials and techniques from each other. William Swainson observed that "lampreys, eels, and other cylindrical fish may be preserved by skinning them from the head to the tail, in the same manner as eels are prepared for cooking." George Humphrey, who had compared flattened fish specimens to drawings, encouraged the use of foodstuffs like pepper to preserve half-specimens of fish on paper. Natural history is, in many ways, the history of food, and food history a form of natural history.[13]

Preparing my own fish offered an opportunity to know it, as an animal—at least in part—and to attempt to know, historically, how someone such as Peck or Gronovius might have wrestled with fishy bodies. But it was likewise an exercise in not knowing. The instructions lacked numerous steps, such as how to prepare the eyes, which I surmised on my own, and many of the physical gestures would not translate well into writing. Through my own trials and errors, I realized the impossibility of fully replicating the cultural context of this process, given how vastly my environment differed from that of someone like Gronovius or Peck. Gronovius let his fish dry near the fire or in the sun depending on the season; I had to put mine in a modern appliance. He let his fish's flesh putrefy to separate it more easily from the skin; I have neighbors who would complain, so I found a different, if more laborious, method of separating skin from flesh. An eighteenth-century naturalist might have enlisted help (or the bulk of the labor) from his wife, a servant, a technician, or someone he had enslaved; I had the aid of a dear friend and equal. As a method of revealing uncodified aspects of remote pasts, reconstruction has no hope—and in my case, no real desire—to repeat history in a literal sense. Its importance lies in the new questions it makes possible, which are much more difficult to ask otherwise. Its significance lies in how performing a procedure might hint at all the other procedures and labors and participants that never made their way into textual accounts.

My inability to know, to reach across time, felt especially poignant for this particular reenactment. Flattening has always entailed its own politics. The very rhetoric and visual logic of flattening erase traces of labor, as well as crucial forms of expertise—such as the skill needed for catching and cutting and cooking fish—that underwrote natural history's practice. The final product seems simple, self-evident, and naturalized. In that way, fish flattened on paper are not all that dissimilar from modern-day infographics, charts, and other two-dimensional visualizations, which Johanna Drucker has likened to an "intellectual Trojan horse" for their claims to transparency, their seemingly unimpeachable presentation of bare facts. Moreover, period instructions for flattening fish well beyond Gronovius overwhelmingly stress its ease. Samuel Latham Mitchill underscored its "facility" in his proposal for fish crowdsourcing discussed in the previous chapter; Swainson called it "the most simple method" for preserving fish; Lettsom reassured that the skin "may be slid off almost like a glove."[14]

But generating these specimens was often anything but simple, all the more so for historical makers without my privileges. For some, it was carried out in the bonds of racial slavery, amid literal tempests, as in the case of the man who furnished Alexander Garden, and thus Linnaeus, with crucial specimens like the striped porkfish. Instructions depict cutting and flattening as unskilled acts, even though getting them right required dexterity and mastery of various craft techniques. These statements echo what Lauren F. Klein, in her study of early American race and eating, has called "the violence that can be enacted through visual display." Reassuring, can-do instructions had a political edge that persisted in the final product, as flattening sutured these animals to paper bearing Latin rather than Indigenous names and set the animals floating in a pool of whiteness, a signal of all that was absent. Even so, those larger histories remain congealed in the specimens.[15]

Eighteenth-century ichthyologists loved to bemoan the neglect of their subfield. Peck's own specimens were reportedly misplaced at Harvard for many decades. In the 1930s, Thomas Barbour, then the director of Harvard's Museum of Comparative Zoology, tried to write a biography of Peck over the course of many years. He eventually gave up, insisting the story "would not pull into a yarn." Barbour described the flattened fish prepared by Peck as "mounted after the crude manner of the day" and elsewhere echoed the labor-belittling undertones of eighteenth-century instructions when he wrote that Peck "was drying out and pressing fish skins and sticking them on cards, making very crude reference specimens." When trying to track down evidence that George Washington had seen Peck's fish and admired their excellence, Barbour sneered that such praise would be "hard to believe if you could look at them now." Barbour's contempt for the specimens belied their technical precision, the value placed on them in the eighteenth century, and their utility and ubiquity in natural history—and thus, modern biology's debt to them. Peter Davis has written: "Skins preserved in this way remain as some of the most important historic and taxonomic collections in ichthyology." Barbour, for his part, concluded that Peck was "a dreary, tiresome letter writer and a very undistinguished diarist" and that most of the materials about Peck's life story that he could amass were "fragmentary" and "absolutely without any human interest"—even if Peck was "an awfully good naturalist." Peck has largely fallen into obscurity, even though he authored what is recognized as America's first article on systematic zoology. The one on fish.[16]

Nevertheless, fish flattening appears to be having a moment. After generating my own specimen, I learned I was not the only researcher in the world with a taste for piscine reenactment: a group of scholars at the University of Leiden also recently replicated the Gronovius technique as part of a larger project on the history of ichthyology, as did a wholly separate group in Portugal after discovering a cache of long-lost flattened specimens at the University of Coimbra. Both teams of reenactors (which also happened to include scientists), like most of my literate historical actors, stressed the accessibility of the technique given the low-tech tools involved, though their documentation of the process intimates the uncertainties and gore along the way—including their explicit recognition of the difficulty of parting the skin and flesh, despite Gronovius's insistence to the contrary.[17]

Why have so many turned to flattened fish now? Curiosities then, and curious now, the specimens are quite simply wonderful and strange. They appear the most inert of the animal traces I study. But at times they also feel the most immediate and alive. You may recall that the first faint intimation of what this book might become entered my mind as I stared in confusion at a dried pickerel from Peck's collection, marked up with numbers that I later learned were fin ray counts (figure 6.4). The resurgence of fish flattening might be tied to do-it-yourself culture, the so-called maker movement, and a recent materialist turn in academia. Yet I think it runs deeper than that. These specimens continue to speak to us because they still have more to say. As Gronovius said, "Paper always sticks."[18]

But I wasn't done with fish just yet. Or perhaps the fish weren't done with me. Purpose had propelled my journey to the fishmonger to flatten eighteenth-century style: I wanted to understand the Gronovius instructions by doing. My

Figure 6.4. William Dandridge Peck, dried pickerel skin mounted on paper, 1790. (Collection of Historical Scientific Instruments, Harvard University, 0039; courtesy of the Collection of Historical Scientific Instruments)

next déjà vu moment, however, would happen much more fortuitously. While ambling through the waterfront district of Salem, Massachusetts, on a summertime weekend excursion, I passed by a sign on a quaint yellow storefront that read: "FISHED IMPRESSIONS PRESENTS: JOE'S FRESH FISH PRINTS: *Printed the old fashioned way,* WITH INK, PAPER, AND A FISH." Who was Joe, I wondered? How and why was he printing fish? Were these engravings or woodcuts like the ones I stared at so often in old natural history volumes, or something else? How "fresh" could a print be?

I walked into the store and soon recognized a technique I had only read about before: *gyotaku.* This Japanese fish printing method, traditionally used by fishers to document their catch, directly transfers the inked skin of fish and other aquatic creatures to paper (figure 6.5). Joe Higgins, the owner, artist, and a lifelong fisherman, wasn't in that day, but I later met him when I signed up for one of the store's fish printing sessions. Under cover of darkness, William and I, alongside two other couples, painted the sides of striped bass with bright blue ink and rubbed pliable rice paper over their bodies to take impressions, our hands acting like a gentle printing press.

Joe's images came out breathtakingly clear and alive at our session. His

Figure 6.5. *Gyotaku* artwork by Joe Higgins.

artworks adorned the walls surrounding our worktable. There were his signature severed tuna tails, two white marlins caught in a standoff with a dancing shoal of squid, and a ruddy octopus brewing its own inky creation as it expelled a black smokescreen. My printing attempts, in contrast, looked muddy at best. But all of our prints were flattened and in profile view, just like fish herbarium specimens. All were likewise distorted by that perennial problem of translating from three to two dimensions: Joe said fish always ended up slightly larger than their actual size when printed. Just like Gronovius and Peck, Joe would spread each fish's fins open with pins to simulate a lifelike pose. Unlike a high-stakes herbarium specimen, however, these prints were reproducible and allowed room for error. One could continue to ink the animal's side over and over, making one print after another. Joe also said he never let a specimen go to waste. He applied lemon juice to remove the ink, leaving the fish intact for consumption. He could even print crabs while they were alive.

Aside from being New Englanders who drew from traditions of artisanal knowledge, Joe Higgins and William Dandridge Peck shared a few other things in common. For both men, different fish species presented different

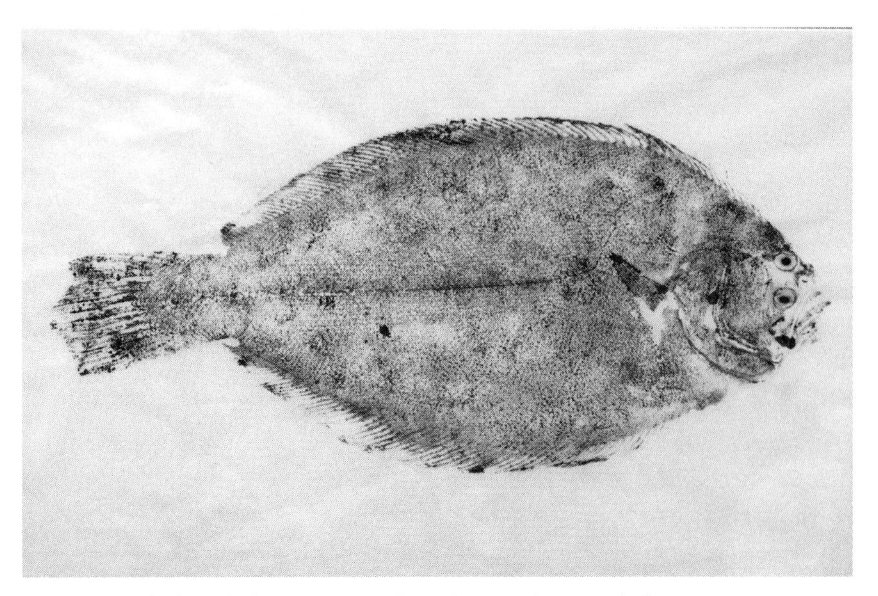

Figure 6.6. A flatfish inked on paper. (Randy Duchaine / Alamy Stock Photo)

opportunities and different pitfalls. When I asked Joe what the simplest fish to print was, he immediately cited a flounder as an example of a straightforward *gyotaku* specimen, given its flattened form (figure 6.6). He was in good company with Peck there. I then asked what species he found most trying. Monkfish had a difficult orientation, Joe said. Skates and stingrays were challenging for similar reasons. Octopuses made Joe sad to print, given their intelligence. Moreover, they were quite slippery to capture as artwork since their grocery store appearance, the state in which he usually acquired them, looked alien compared to their fresh-caught vibrance.

Out of curiosity, I showed Joe a photograph of Peck's deflated lumpfish specimen on my phone. After a heavy pause, Joe said ominously, "That fish does not *want* to be printed"—over and over, almost as if to console himself. He went on to reveal that he had tried to print a lumpfish fifty times or more. And it never worked. Cloth offered a better substrate than paper in this case, so he had some success translating a lumpfish's likeness to a t-shirt. But Joe still wasn't satisfied. Part of the issue seemed to be textural. Lumpfish lack scales, instead having smooth skin that might be covered in bony tubercles depending on the species. Part of the problem also seemed dimensional. The lumpfish's boldly round body, which dramatically tapers into a steep moun-

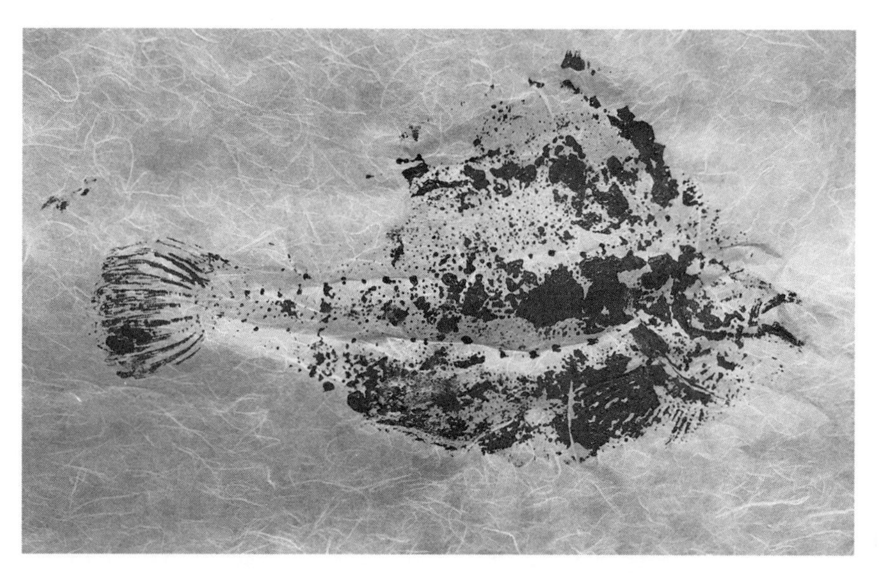

Figure 6.7. A lumpfish eluding the page. (Photograph by Whitney Barlow Robles)

tain ridge at the top, made for an especially awkward fit on a flat surface. This fish had been bugging Joe for some time, it seemed. He happened to have one stored in the store freezer, which he offered to me for experimentation. If dexterous Joe had struggled, I obviously didn't stand a chance. My attempted prints are full of white patches: they are ghost images, the fish having escaped the net of the paper (figure 6.7). This particular animal resists flattened representation even today. Joe, for his part, said I should take the lumpfish home, having resigned himself to the creature's elusive nature.

One of natural history's most intractable dilemmas in the eighteenth century revolved around capturing ephemeral life-forms for distant readers, patrons, and fellow naturalists: the challenge of putting life on a page. One solution was to do just that, affixing bodies to paper. As fish herbarium specimens reveal, and as my trials flattening fish and inking them onto paper as prints also confirm, the animals themselves determined the viability of such strategies in species-specific ways. For the trickiest types, fish had the last word.

RACCOONS

In gentleness of heart with gentle hand
Touch,—for there is a Spirit in the woods.

—*William Wordsworth, "Nutting"*

CHAPTER SEVEN

Sleight of Hand

YOU DRAG YOUR HANDS OVER THE SAND, across the slime, until you feel a lump. Water passes through your fingers—the more, the better. A grooved crescent object takes shape in your hand and then in your mind. It is an object you have felt many times before. Your palms explore it, your digits find the crease, and with a firm tug the thing splits apart. A pearl inside to whet the appetites of jewelers: of little value. The viscous mass beneath it: a meal. Your head and your hands will remember this spot.

This scene, played over throughout your life, would prompt naturalists like John James Audubon and John Bachman to rank your knowledge above that of a human scientist. Though people could never actually experience what it is like to be you, your perspective and your perception make natural history look different. For they can open a historian's eyes to nonhuman ways of knowing the world—to animal forms of expertise on the margins of the Enlightenment project. The impossibility of ever fully imagining or inhabiting your position makes the thought experiment all the more necessary. I must make you, dear reader, a raccoon.[1]

As humans, we gather information about our environment through a combination of vision, hearing, smell, taste, and touch. Raccoons likewise possess these senses, but they understand their worlds conspicuously through their hand-like paws. Raccoons' sensitive palms harbor an unusually high number of sensory receptors, while the portion of their brain that processes tactile forepaw information trumps that of most other mammals in its size and complexity. Water, moreover, appears to enhance this manual fact finding. As raccoons grasp the surfaces around them, running their paws across intertidal sands to find clams and oysters or rolling objects in their palms to understand their shape, they create tactile topographic mind maps of their surroundings. The raccoon's touch, then, is a sense at once familiar to us, yet also alien. Constance Classen says of touch: "Like the air we breathe, it has been taken for granted as a fundamental fact of life, a medium for the production of meaningful acts, rather than meaningful in itself." Touch slips through our fingers like water. While scholars like Classen have started to historicize touch, uncovering its importance in unexpected corners of the human past, it was and remains the ultimate source of meaning for raccoons. Raccoon hands are a world-building apparatus, with consequences for the human world that studies them.[2]

Hands smack of human exceptionalism, even though many animals—raccoons, apes, monkeys, opossums, and squirrels, to name a few—have anatomically analogous appendages that perform hand-like functions. In one classic account, the philosopher Martin Heidegger linked hands with thought and signification, and he granted them to humans alone. "Apes, too," wrote Heidegger, "have organs that can grasp, but they do not have hands. The hand is infinitely different from all grasping organs—paws, claws, or fangs—different by an abyss of essence." Unraveling the politics of the hand, the schisms between cerebral savants and manual laborers, has spawned something of a cottage industry among historians. Steven Shapin and Barry Barnes, for example, have shown early modern English attitudes toward the head as "reflective, manipulative and controlling; the hand, unreflective, mechanical, determined by instructions." Touch often formed the bottom of the sensory hierarchy in Western thought. Yet these literatures on hands and touch remain resolutely human. If hands historically played a role in attempts to differentiate people, they also had a hand in policing the boundary between humans and animals.[3]

Raccoons have always had a lot going for them. They bear several humanoid features, like forward-facing eyes, masks, and the ability to walk on two legs. (Bipedalism was yet another feature said to literally elevate humanity above the brute animals.) Raccoons also sport wholly unhuman yet utterly remarkable ringed tails. In early America—raccoons were once restricted to the Americas, before conquering a wider geography—they were economically valuable and practically useful animals. Markets prized raccoon skin second only to beaver.

Its warm embrace became everything from coonskin hats to "fine Womens Shooes," while colonial travelers converted the interior "grease" of raccoon bodies into medicine and their flesh into fare. Nevertheless, it is their hands that have occupied the human gaze for centuries. The word "raccoon" comes from the Algonquian *arakun* or *arakunem,* meaning "it scratches with its hands." By the time European conquerors arrived in the Americas, raccoons had long pervaded Native communities as utilitarian tools, symbolic objects, and companions. Over the next few centuries, raccoons would beguile the scientific imagination of naturalists from abroad, enter menageries as captives and homes as pets, and, along with various other animals native to the Americas, embody the newness of the so-called New World through their unique habits and the challenge they presented to efforts to pigeonhole them into traditional taxonomic hierarchies.[4]

The manual dexterity of raccoons, and especially the appearance that they washed their food as they manipulated objects in water, became a chief reason they were studied and brought into homes in the first place. This behavior would become enshrined in the many vernacular and scientific names given to raccoons; even the fingers of human language, as we will see, interlaced with the raccoon's hand. A raccoon's way of interacting with its world, driven by the creature's much-celebrated intelligence and inquisitive nature, in turn shaped how people approached a raccoon's world with their own curious agenda. Raccoon forepaws became a bridge, if ever a shaky one, between humans and animals.

Raccoon curiosity helped establish a growing recognition that thinking minds, or knowing and probing hands, might not exist solely among humans and hence might not separate humans from animals. Although raccoons had no intentions of using their curiosity to classify natural objects according to human rubrics like the Linnaean system—nor could they—raccoons do, for instance, use touch to distinguish species of intertidal clams to maximize energy profits. They need not even use their eyes for these sensitive acts of sorting. Raccoons classify objects and create taxonomies, though they are taxonomies based on raccoon priorities alone. They make, and apply, knowledge.[5]

When considering our contrasts and affinities with other animals, we are ever prone to navel-gazing, searching for their possession or want of things that seem especially human and thus worthy to us—traits like reason, logic, language. Some eighteenth-century naturalists genuinely believed that certain animals could employ something akin to reason or pull off other intellectual feats. Visions of the Chain of Being placed especially knowing or powerful animals just a shade below man—like the monkey, dog, and lion in the rendering that accompanied the work of Charles Bonnet, as we saw in the introduction. Joseph Banks put humans at the summit, but right beneath them lurked "the half reasoning Elephant, the sagacious dog, the architect Beaver, &c." Raccoons don't typically appear in period illustrations of the Chain of Being, though the inner mental workings of raccoons generated substantial interest from natural historical writers and, as intelligent mammals, they would have occupied its loftier animal heights.[6]

But in the face of their explicit absence, what might it mean to contemplate the affinity of human and animal worlds in a way that doesn't merely revolve around reason and cold rationality—or around us? What if we oriented our investigation around curiosity—around asking questions rather than giving answers? What if instead of looking for the human in animals, we looked for the animal—the manual, the material, the appetitive—in our own systems of knowledge? What if we turned the Chain of Being on its head? For one could make knowledge without necessarily using reason. And raccoon curiosity ends up resembling the naturalists featured throughout the pages of this book quite well: raccoons and naturalists were driven by survival instincts, the gut, a proclivity to eat their subjects of study, and a desire to get their hands on new objects. In the words of the Welsh naturalist Thomas Pennant, a raccoon was a busy investigator, "being always in motion, being very inquisitive, and examining every thing it sees with its paws." Considering such alien curiosity, this touchy-feely underbelly of raccoon knowledge, sheds fuller light on the multisensory apparatus of human knowledge at the same as it is also not fully coterminous with our knowledge.[7]

Naturalists ultimately tried to understand and stabilize the raccoon's knowledge and new nature through taming the animal, converting pet-keeping into a domestic laboratory. However, the ever clever raccoons confined in yards, parlors, and menageries frequently scrambled the agendas of naturalists. They thwarted experiments, escaped from captivity, and generally did as they pleased. Their mixed ability both to accept something like domestication while also escaping it reinforced the idea of the New World as truly unruly and new. It put raccoons' full apprehension out of naturalists' grasp. The knowledge of humans and the knowledge of raccoons existed in reciprocal tension: people wanted to understand raccoon curiosity, but the curiosity of raccoons kept them from being fully understood. Figuratively and literally, raccoons refused to sit quietly where humans put them.

PLACING THE RACCOON

When James Parsons—the same London physician who was fascinated by polyps—encountered a *Siyah-ghush,* or caracal, for the first time, he realized some comparisons were in order (figure 7.1). Today a textbook or encyclopedia might describe the caracal as an established beast: a tawny wild cat found in Asia, the Middle East, and Africa that is strikingly identifiable by comically long black tufts that protrude from its ears. But in 1760 Parsons sought to make sense of the unfamiliar nature of a live caracal brought to London. He found the animal had a "head like that of a cat," "paws of a cat," and behaviors "like those of a cat." It licked its foot and drew it over its face "exactly like a cat," hissed when harassed, and had teeth "in the same number and manner with those of a cat," while its body terminated in a long and slender "tail like that of a cat." Other naturalist authorities, Parsons noted, ranked it among the cats. It looked like a cat, smelled like a cat, talked like a cat, and walked like a cat. From the foregoing, Parsons felt somewhat confident stating: "I am inclined to rank this animal among the cats."[8]

Raccoons weren't so simple. As naturalists and travelers encountered strange beasts in the early modern period, they operated within vocabularies of animals familiar to them. And where cats and many other animals were concerned, comparison by analogy could serve them well. In the New World, squirrels, foxes, wolves, bears, beavers, and even ancient mastodon bones evoked (and were usually related to) their Old World analogues. The uniquely American raccoon, however, remained an enigma, for it lacked a European comparand.

That seldom stopped naturalists from grouping it in existing categories, however. Raccoons conjured a range of animal analogies, most commonly of bears and monkeys but also canines, badgers, rats, and—as Parsons might approve—felines. For some, a single comparison stood out. The Reverend John Clayton branded the raccoons of Virginia a "species of a monkey," focusing on their "feet formed like a hand," monkey-like face, and "very apish" qualities when kept tame. John Ray gave the animal one of its earliest Latin names by describing it in 1693 as *Vulpi affinis Americana,* thus noting its similarity to foxes (which are now grouped under the genus *Vulpes*). Ray's observation held

Figure 7.1. Figure of a caracal from James Parsons, "Some Account of the Animal Sent from the East Indies . . . ," *Philosophical Transactions* 51 (1760). (Royal Society of London)

true decades later for the English naturalist Mark Catesby, who likely observed raccoons on the ground in America during his travels there. Catesby listed raccoons under the heading "Beasts of a different Genus from any known in the Old World" and remarked that they "resemble a Fox more than any other Creature, both in Shape and Subtlety, but differ from him in their Manner of Feeding, which is like that of a Squirrel." And Edward Tyson, who performed extensive anatomical studies of New World oddities like rattlesnakes and opossums at the Royal Society, suggested in 1704 "a New Division of Terrestrial Brute Animals, particularly of those that have their Feet formed like Hands," and who possessed "*Fingers* rather than *Toes*." Tyson believed that some such animals should be designated *Quadrumanous,* or four-handed, instead of the more typical period designation of quadruped, meaning four-footed. He placed raccoons and coatis, close ring-tailed relatives of raccoons, alongside monkeys and apes in this system, not due to similarities in overall bodily architecture but due to the shared feature of hands.[9]

Other naturalists and explorers, however, found these beings still less determinate and described the raccoon as a composite of familiar animals instead. In the 1730s, the British writer and traveler Francis Moore voyaged to colonial Georgia and regarded the raccoons he observed to be "something like a Badger, but somewhat less, with a bushy Tail like a Squirrel, tabbied with Rings of brown and black." These chimerical characterizations persisted into the nineteenth century and beyond, for American as well as European eyes. In 1842, for example, the American zoologist James E. De Kay said the raccoon had been "quaintly described as having the limbs of a bear, the body of a badger, the head of a fox, the nose of a dog, the tail of a cat, and sharp claws by which it climbs trees like a monkey."[10]

Those sharp claws and especially the hands to which they were attached drew the attention of almost every naturalist who encountered raccoons and even informed their formal classification. In theory, systematic classification enabled naturalists to impose order and hierarchies on the natural world. Through sorting and naming, they could understand, manage, and bridle nature's complexity—what Harriet Ritvo has called an attempted "appropriation and mastery of the animal kingdom." Since the late eighteenth century, it had been said of the Linnaean project: "God created, Linnaeus set in order." Filing raccoons in existing classifications or among groups of well-known animals represented one effort to maintain sovereign control over the proliferating animal categories that resulted from the exploration of the Americas and beyond.[11]

Linnaeus, for one, classified the raccoon as a type of bear. No doubt many others named it as such by following his lead or through independent observation of its bear-like habits and morphology, which often focused on the structure and activities of the animals' feet. Linnaeus originally named the raccoon *Ursus cauda elongata,* which translates to "long-tailed bear." He changed the raccoon's Latin name to *Ursus lotor* when he converted to a binomial system of nomenclature with the 10th edition of his *Systema Naturae,* published in 1758—an edition that would also herald the creation of categories like *Primates* and *Mammalia. Ursus lotor,* roughly meaning "washing bear," foregrounded the raccoon's behavior of manipulating its food and other objects in water.[12]

In fact, naming and classification conventions for raccoons, formal and otherwise, quite often drew explicit attention both to the animal's hands and the hand's role in their curious behaviors. Water appears to enhance the sensitivity of nerve endings in raccoon paws, thus magnifying the animals' tactile understanding of objects in their grasp. Among Europeans, the raccoon's penchant for rolling and dunking objects in water—its reputation as a "washing" animal—would also be enshrined in the German word for raccoon (*Waschbär*), the Spanish word for the crab-eating raccoon (*osito lavador*), one early French term for the animal (*raton laveur*), and the modern Swedish word for raccoon (*tvättbjörn*). Linnaeus's broader genus *Ursus*, which included wolverines and badgers alongside raccoons and true bears, grouped these animals according to their shared dental and ocular features, the smoothness of their tongues, the prominence of their snouts, their curved penis bones or baculums, and their plantigrade manner of standing—the last of which also focused on the creatures' feet. (In contrast to digitigrade animals like cats and dogs, which actually stand on their toes—in other words, what might look like a cat's elbow is really its wrist or ankle—plantigrade animals like bears, humans, and raccoons walk on the soles and heels of their feet.) In all of these ways, raccoons' tactile curiosity infiltrated human attempts to place them within Enlightenment frameworks rather than contending with the radical difference they presented.[13]

Linnaeus, for his part, learned something about raccoons by reading other authors. But he also possessed firsthand knowledge. Through Linnaeus, we have something that is often lacking in the historical record for corals, fish, and even rattlesnakes like I.II.I.a: we have a name. Linnaeus kept a number of pet-subjects at the botanic garden of Uppsala University, including a West African monkey named Diana who delighted in raisins and provided Linnaeus with the name for the Diana monkey, *Cercopithecus diana*. Crown Prince Adolf Fredrik gave Linnaeus a raccoon, named Sjupp, used here by Linnaeus as a proper name, though it derived from a generic Swedish term for raccoons and their pelts. Six months later, in 1747, Linnaeus published a lively character study of his pet, this "rare quadruped animal," in the proceedings of Sweden's Royal Academy of Sciences.[14]

Linnaeus noted Sjupp's acute sense of smell but said he was hard of hearing and "completely blind." It's possible Sjupp suffered an illness or some trauma on his voyage to Europe. Linnaeus focused on his ursine traits, calling Sjupp "our American Bear," and lingered on the centrality of his hands in sensing the world. "This bear had a sense of touch as acute as that of any animal," wrote Linnaeus, "so that he could find whatever was thrown to him even though he couldn't see it; by patting the ground with his soft hands he could locate the smallest crumb." Despite Linnaeus's rapt attention for Sjupp's remarkable penis bone, the Swede described Sjupp in feminized terms, drawing attention to his delicate forepaws and calling their palms "wrinkled and black, but soft as a Virgin hand." Linnaeus also admired how Sjupp formed his hands into the shape of a spoon for taking liquids. Sjupp even had a taste for rifling through unsuspecting visitors' pockets for hidden morsels, pockets that he manually "scrutinized" in search of "contraband," which the raccoon "violently forced . . . out." Sjupp would then hold fast to the victim's clothing, biting anyone who tried to extricate him. Using similes much as Parsons would with his caracal, Linnaeus belabored the bear-like qualities of Sjupp in noting that he took things into his hands like a bear, "walked just like a bear, sauntering on his heels, his long thighs and legs held widely apart, his back bowed and his head lowered; yet when he noticed something that pleased him he could walk quite a long way on two legs, erect like a bear."[15]

But then, tragedy befell our American bear. Sjupp broke loose one night—a running theme among captive raccoons—and met his end with a large dog after entering a neighboring yard. It took several days to locate the body. Sjupp's misfortune, however, was natural history's boon. In the depths of "great sorrow," Linnaeus promptly dissected and described Sjupp, his former friend and beloved pet, as a specimen. After death, Sjupp became the skin of his namesake. He lives on through an illustration that accompanied Linnaeus's published study. A magnified depiction of Sjupp's penis bone floats in the foreground, while Sjupp's hands are highly stylized (figure 7.2). His hands also figure prominently in an emotive pen and ink drawing on which the printed engraving seems to be based (figure 7.3). And Sjupp wasn't the only live raccoon model who would inspire a European artist to document its hands in sharp detail. In the previous century, the Flemish painter Pieter Boel depicted an array of live subjects from King Louis XIV's royal menagerie, including two images of a raccoon on a canvas shared with several studies of a marmot (figure 7.4). In one rendering, the raccoon looks squarely at the viewer with its forward-facing eyes and black mask, its finely delineated hands planted on the ground. In the other, the raccoon reaches a paw toward the edge of the canvas, evoking Michelangelo's painting *The Creation of Adam* on the ceiling of the Sistine Chapel—an image whose contours have been likened to the human brain.[16]

For centuries before Tyson, Linnaeus, and Boel, however, Native peoples

Figure 7.2. Engraving of Linnaeus's pet raccoon, Sjupp. The bone on the bottom is the animal's baculum, or penis bone. (From Carl Linnaeus, "Beskrifning På et Americanskt Diur, som Hans Konglige Höghet gifvit til undersökning," *Kongl. Svenska Vetenskaps Akademiens Handlingar* 8 [1747]; Biodiversity Heritage Library)

Figure 7.3. This pen and ink drawing of Sjupp, attributed to Laurentius Alstrin, likely served as the basis for the engraving printed with Linnaeus's account. Pin pricks in the paper suggest that Linnaeus may have hung this drawing on the wall. (Laurentius Alstrin, "Drawing of a Raccoon," c. 1746, GB-110/LM/PF/ALS/1; by permission of the Linnean Society of London)

Figure 7.4. Detail from Pieter Boel, *Etudes d'une marmotte et d'un raton laveur,* c. 1670; Musée des Beaux-Arts de Bordeaux, Bx E 915. (Erich Lessing/Art Resource, New York)

across the Americas had already centered the raccoon's hand in their names and descriptions of the animals. Western science once again found itself playing catch-up, as it had in recognizing the sociality of rattlesnakes. In addition to the previously mentioned *arakun,* variants of the Anishinaabe word *esiban* (meaning "it picks up things") can be found in dozens of Native languages. The word for raccoon in Chickasaw (*shawi'*) and Choctaw (*shaui*) translates to "grasper." The Nahuatl (Aztec) word *mapachitli,* which provided the basis for the Spanish word for raccoon (*mapache*), means "they take everything in their hands." Several Native traditions also preempted Linnaeus in classifying raccoons as the kin of bears.[17]

While the worldviews of Native Americans were as diverse as those of Europeans, Native societies generally embraced more capacious notions of kinship than did Europeans, including acknowledging human ancestry with animals. In the words of the anthropologist Laura A. Ogden and the ecologist Nicholas J. Reo, a citizen of the Sault Ste. Marie Tribe of Chippewa Indians, "Not only are plants and animals people, but they are kin, or part of Anishnaabe extended family." This offered a different and more fluid baseline for assessing an animal's humanoid traits. The Chickasaw, Muscogee Creek, and Menominee, among others, each have a Raccoon Clan. In various lines of Native thought, animal and people can morph into one another, as in the case of raccoon "man-beings" who switch among raccoon, rattlesnake, and human forms in Wyandot and other traditions.[18]

Indeed, as Europeans catalogued the biodiversity of the Americas and encountered new-to-them animals, they worked in the shadow of long-established Native knowledge. The humanish traits of raccoons, especially the remarkable ways in which they marked up the world with their hands, likewise figured in Indigenous classifications. One especially rich natural history of raccoons appears in the Florentine Codex, an encyclopedic compendium of Mesoamerican thought that represents the fruits of a collaboration between the friar Bernardino de Sahagún and Nahua scholars and artists in the sixteenth century. Spanish writing, Nahuatl descriptions, and arresting images intermingle on each page. In a section on raccoons in Book 11, titled "Earthly Things," humanoid hands become the distinguishing feature of the animals. The text explains that the raccoon received its Aztec name, *mapachitli,* "because its hands are quite like our hands"—an observation stated five times in the space of three short paragraphs. The accompanying illustration shows a gray, almost wolf-like beast (figure 7.5). Though it lacks the distinguished mask or ringed tail of a raccoon, it boasts unmistakable hands and fingers. The narration proceeds to project this hand-centric characterization back onto humans, noting that, since the raccoon "takes everything that it sees, and since it has human hands, a thief is also called *mapachitli.*"[19]

Francisco Hernández, a physician to King Philip II of Spain, relied on both Sahagún's writings and Native informants to compile a lengthy natural history of New Spain when he explored present-day Mexico from 1571 to 1577. Here again, the hand reared its head. A section on the quadrupeds of New Spain opened with the *mápach,* or raccoon—a section Hernández subtitled "the animal that feels everything with its hands." Hernández noted the tendency of others to shoehorn raccoons into existing groups like apes or foxes, but he believed it to be "a unique animal, distinct in form and habits from all the others I have known." Hernández described the raccoon's frequent presence in Indigenous homes, noted its propensity to steal, and stressed its "manos de hombre," or hands of man, which it used to interrogate everything in its reach.[20]

Figure 7.5. Image of a raccoon or *mapachitli*, from Fray Bernardino de Sahagún, *Florentine Codex: General History of the Things of New Spain*, Book 11, "On Earthly Things." (Florence, Biblioteca Medicea Laurenziana, Ms. Med. Palat. 220, f. 164r; by permission of the Ministero della Cultura; further reproduction is prohibited)

European and Indigenous classifications of raccoons could be volatile and ever shifting, not unlike the beast itself. This was heightened by raccoons' kinship with other New World animals. Today biologists group raccoons and their closest relatives—coatis (or coatimundis), ringtails, cacomistles, kinkajous, olingos, and olinguitos—into a single shared scientific family called Procyonidae. In the early modern period, naturalists often mistook raccoons for South and Central American coatis and vice versa, or simply used the terms "raccoon" and "coati" interchangeably. In the words of the Florentine Codex, a coati, or *peçotli*, is "just like the raccoon, but it does not have human hands." Coatis are long-snouted mammals that boast a ringed tail similar to that of raccoons. Unlike raccoons, they hold their long tails to the sky like cats, making their boisterous packs resemble small armed battalions. The animal labeled "coati" in Ole Worm's *Museum Wormianum* (1655), replete with engravings and woodcuts of specimens from Worm's famed cabinet of curiosities, most likely depicts a raccoon rather than a coati given its shorter snout and other features (figure 7.6).[21]

Some naturalists lacked access to live specimens or even dead ones, having only the words of other naturalists as a basis for their descriptions of exotic animals. But firsthand observation allowed for plenty of confusion as well in the face of the strange and variegated quadrupeds of the New World. A man named Joshua Spencer drew a raccoon from the flesh in his commonplace

Figure 7.6. Engraving of an animal labeled "coati"—almost certainly representing a raccoon rather than a coati, given its countenance —from the collection of the Danish naturalist Ole Worm. (Ole Worm, *Museum Wormianum* [Leiden, 1655], 319; Biodiversity Heritage Library)

book, calling it "a Rackoon or Coati" and "The West India or American Fox." He claimed to have viewed several live individuals exhibited in England in the first decades of the eighteenth century (figure 7.7). Beneath the raccoon, he drew what he called a "Picary or Coati Mondi," which he also observed in person. Despite a coati's genuinely piggish snout, the drawing still manages to resemble a monstrous hybrid of a coati's body—including clawed feet and a ringed tail—with the head of a peccary, wild relatives of swine, for whom "picary" was a variant eighteenth-century spelling. But Spencer's chimera actually subtly mirrored Amerindian perspectives. As described in the Florentine Codex, the Nahuatl word for coati—*peçotli*—could refer to any person or beast who eats voraciously, explicitly including both coatis and peccaries as beastly bearers of the term. New World animals, then, had the capacity to shapeshift in the eye of the beholder.[22]

In the late eighteenth century, raccoons would receive their own formal taxonomic lair freed from the likes of bears, apes, or foxes, though not without some canine undertones. In 1780, Gottlieb Conrad Christian Storr devised the genus *Procyon* (roughly translating to "before the dog"), which biologists use to this day to describe the three recognized species of raccoons. But their tactile manner of sensing the world would stick in scientific parlance. The so-called

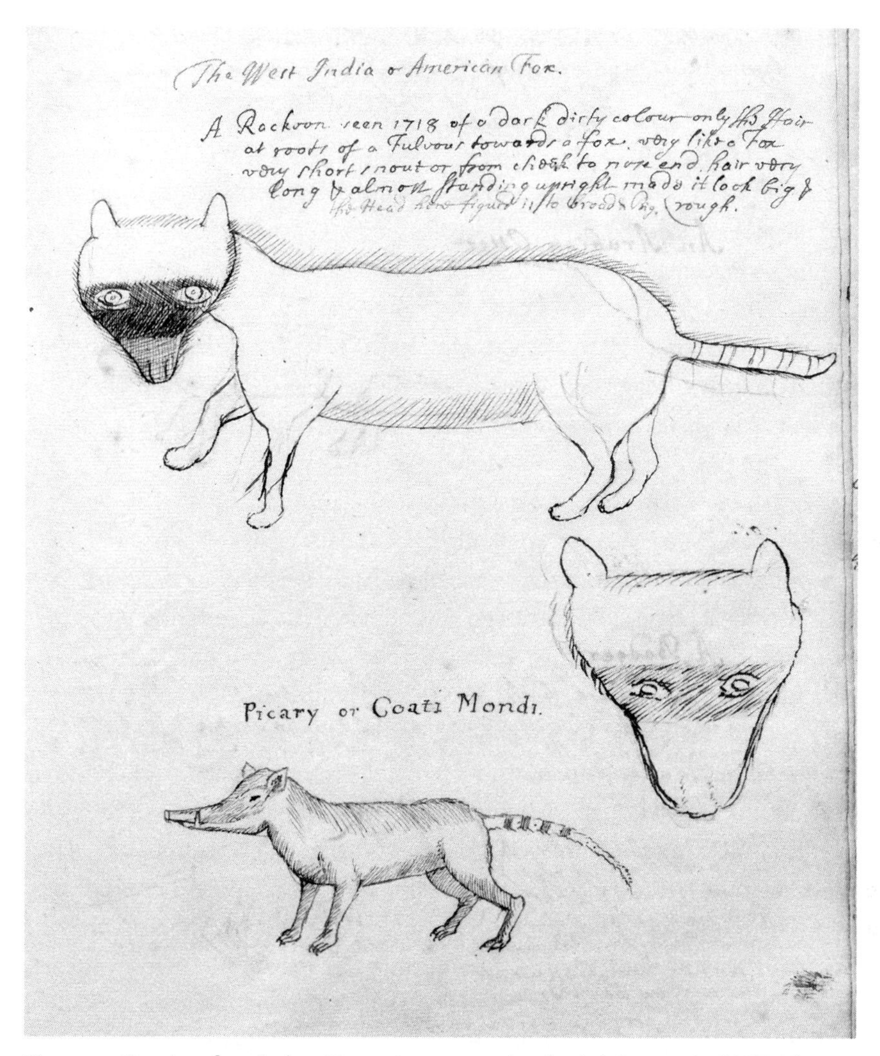

Figure 7.7. Drawings from Joshua Spencer's commonplace book (1692–1729). (ANSP Library & Archives Collection 426)

common raccoon familiar to most, the predominant subject of study herein, still retains its species name from Linnaeus as *Procyon lotor,* while its relatives, the crab-eating raccoon (*Procyon cancrivorus*) of Central and South America and the critically endangered Cozumel or pygmy raccoon (*Procyon pygmaeus*) of Cozumel, Mexico, have been granted their own species names. For the most notorious and widespread species of this genus, then, the hand has persisted lexically. The way raccoons came to *know* their own worlds helped mold human perception and knowledge of what a raccoon *was.*

HEAD AND HAND

Those forepaws that guided the raccoon's classification through its many twists and turns also made it part of a cohort of animals whose seemingly anthropoid features suggested commonalities with humans, and perhaps even human minds. Although raccoons were by no means regarded as the most humanlike animals—monkeys and apes usually took that prize—they became uniquely bound up with questions of animal knowledge and its complex relationship to the body. Hands became points of tension as to what it meant to be a curious species, human or otherwise.

In her book *The Body of the Artisan,* Pamela Smith argues that the early modern period saw the erosion of a hard boundary between theory and practice, or head and hand. Naturalists and other learned practitioners summoned their corporeal bodies to understand natural phenomena, rejecting the earlier medieval emphasis on insular scholastic book learning. Enlightenment-era discourse may have seen a *rhetorical* emphasis on distanced visuality, framing the eye as the supreme bodily conduit of natural historical knowledge. And some students of early natural history have followed that rhetoric; Michel Foucault once wrote, in a nod to Denis Diderot, that "the blind man in the eighteenth century can perfectly well be a geometrician, but he cannot be a naturalist." However, more recent scholars have demonstrated the importance of non-visual touch, smell, taste, hearing, and bodily knowledge of nature and matter in scientific and medical investigation. People on the margins of science, such as artisans and craftspeople, vitally expanded natural knowledge, even if some eighteenth-century naturalists and philosophers would hold their expertise at arm's length in hopes of guarding their own authority. But additional liminal figures—nonhuman ones—also found themselves enrolled in these vexing debates over the relationship between body and mind and the role of each in creating knowledge.[23]

Naturalists and philosophers disagreed as to whether the presence of analogous body parts in animals and humans implied continuity between those worlds. Some took to the human hand as a means of distinguishing humans from other animals, focusing on its finely tuned structure. This debate raged across the long eighteenth century. As late as 1833, Sir Charles Bell wrote a tract on the hand as part of the Bridgewater Treatises, a series of arguments for evidence of God's design in nature sponsored by the Royal Society at the bequest and dying wish of Francis Henry Egerton, eighth Earl of Bridgewater. Bell noted humanity's seeming ability to conquer "every climate" and argued that the human hand, paired with the human mind, gave people "universal dominion." Bell also cited John Ray, an intellectual forebear of the Treatises. In his influential work *The Wisdom of God Manifested in the Works of the Creation* (1691), Ray dwelled on the "superlative Instrument" of the human hand. Though he admitted that humans lacked the serviceable claws and spurs and shells of other animals, a hand, he insisted, "with Reason to use it, supplies the Uses of all these, that's both a Horn and a Hoof, a Talon and a Tusk, &c. because it enables us to use Weapons of these and other Fashions, as Swords, and Spears, and Guns." Hands turned man into an imperial armamentarium, a military beast to best all others. For some thinkers, animal bodies revealed the wisdom of God through their construction, but they did not threaten to depose man from his superior position.[24]

Yet the issue would not be put to rest so easily, in part because of the very architecture holding up the Chain of Being. Georges-Louis Leclerc, the Comte de Buffon, was quite comfortable with blurred lines lower down on the hierarchy of existence. He called a polyp "the last of animals, and the first of plants" and maintained "there is no absolute and essential distinction between the animal and vegetable kingdoms." But he took great pains to unyoke humanity from the animal world. Buffon struggled with a conundrum that many naturalists steeped in the logic of the chain faced: if its various elements were connected in a unified creation by imperceptibly fine gradations, how could humans, in the face of this divine sfumato, insist on their separation from baser animals? Would they not bleed and blend into the rest?[25]

Buffon nevertheless avowed that humans were "far superior," chiefly because he believed they alone could use reason to doubt, deliberate, and compare. Departing from the more subtle and mixed view of someone like Joseph Banks, Buffon took an uncompromising approach by insisting that just "as there is no intermediate point between a positive and a negative, between a rational and an irrational animal, it is evident that man's nature is entirely different from that of the animal." To those who would ascribe advanced intellectual faculties to other creatures, he asked with some exasperation and brutish territorialism: "Why unnecessarily degrade the human species?" Buffon ultimately argued for the "immateriality" of the human—and only the human—mind, keeping with a tradition that extended back to Aristotle. Unlike the transition from polyp to plant, the leap from person to beast for Buffon jumped abruptly from "a thinking being to a material one, from intellectual faculties to mechanical powers, from order and design to blind impulse, from reflection and choice to ungovernable appetite."[26]

Yet even Buffon, a staunch believer in humanity's singularity, couldn't help but notice the hand's centrality in creating knowledge—even for animals. "It is by the sense of feeling alone," he wrote, "that we acquire real knowledge." For other senses could deceive. As a thought experiment on the importance of touch, Buffon asked his reader to try to inhabit the perspective of a fully grown man who suddenly came into existence with a genuine tabula rasa. Should such a person rely on sight alone, he would open his eyes for the first time and think the grassy hills, the glimmers of light, the blue waters and other distant objects part of his own being. Should he use only his ears, he would think the singing birds trilled from his own consciousness. It would not be until he took his first step and felt this world with his hands and feet and saw it spin around him in relative motions that he would discover its true nature and the bounds of himself. He would reach for the sun, only to find he couldn't grasp it.[27]

Through Buffon's meditations on hands and touch, we can see that even someone committed to the unique reasonableness of humanity might still observe other forms of knowledge-making on the edges of natural history. Specifically, Buffon thought that hands granted knowledge of one's environment by increasing the surface area capable of sensation. A hand divided into flexible and willfully activated fingers, as opposed to a blunt paw or hoof, provided more data points regarding the character of objects in the world. "Those animals which are furnished with hands appear to have most sagacity," Buffon admitted. Within humanity, he believed that women, "among other advantages over the men, have a finer skin, and a more delicate perception of feeling." Touch's power also explained why Buffon opposed swaddling babies, an act he believed deprived their brand-new hands from exploring their brand-new world. (Clearly the Count had never been tasked with soothing a newborn beguiled by the startle reflex night after sleepless night.)[28]

Shuffling back down the Chain of Being, Buffon thus suggested that those "animals which are furnished with many instruments of feeling, as the cuttlefish, the polypus, and other insects, have a superior faculty of distinguishing and of choosing." But in his mind, this meant that fish, whose flat bodies were covered in scales, "ought to be the most stupid of animals, because they can have no knowledge of the form of objects," lacking "divided extremities." He found serpents, however, "less stupid than fishes," for despite their scaly texture, they could wrap and twist their elongated bodies around objects to sense their nature. Buffon might have denied animals rationality and reason, and "thought and reflection," but he nevertheless framed their tactile touch as knowledge and lack thereof as ignorance. Animal knowledge, reasoned or not, sat uneasily within the natural historical project as a sort of irritant—something undeniably present, dinning and buzzing, but difficult to square with the perceived preeminence of humans.[29]

By the late eighteenth century, some would deem Buffon to be on the conservative end of the spectrum regarding animal intelligence, including the capacity to reason or reflect. William Bartram delivered one trenchant critique. He figured that humans dismissed animal ingenuity as instinct because they were too proud to "confess it to be Reason & Intelligence, because forsooth it will detract from the *Dignity of Human Nature*. But where is the proff [proof] of this Dignity of our Nature[?]" Bartram observed that many animals had something of language, tutored their young, and showed "the faculty & powers, of improving, altering or modifying their Maners & arts" in response to changing circumstances. They could also be taught by humans, a fact indicating "Understanding, i.e. The Power of Reasoning or deduction, by a wise perfect comparison & arrangement of Ideas, or notions." Even William Smellie, an Edinburgh naturalist and popular English-language translator of Buffon, would beg to differ with his beloved Buffon in his own writing. In *The Philosophy of Natural History*, Smellie lambasted Buffon's "peculiarly severe" assessment of animal minds, especially when it came to Buffon's dim view of bee intelligence. "Though no animal," wrote Smellie, "is endowed with mental powers equal to those of man, yet there is not a faculty of the human mind, but evident marks of its existence are to be found in particular animals. Senses, memory, imagination, the principle of imitation, curiosity, cunning, ingenuity, devotion, or respect for superiors, gratitude, are all discoverable in the brute creation." Smellie believed animals had their own arts, their own languages, even their own ways of asking for help and lending it to others.[30]

Smellie's worldview was still anthropocentric, as he believed that "the intellect, or sagacity, of inferior animals augments or diminishes in proportion as the formation of their bodies approaches to, or recedes from, that of man." But while Ray, Bell, and their ilk elevated the hand as a marker of human uniqueness, other naturalists would correlate animal hands and animal acumen.

For instance, the same Benjamin Smith Barton beguiled by rattlesnakes also studied opossums. In an attempt to defend them from Buffon, who called the creatures stupid—Buffon became a special target of American naturalists given his claims that New World animals were degenerate—Barton asked: "Would not one be inclined to suspect, from the circumstance of his possessing a *hand,* that the opossum is an animal of at least an ordinary share of intelligence?"[31]

Against this backdrop of debates about animal bodies, hands, and minds, raccoons stood out for their curiosity and cunning and for their entertaining—though at times disconcerting—manual tendencies. Their hands weren't quite as human as an ape's, and they didn't have the renowned sagacity of an elephant. But an emerging research program on raccoons suggested the centrality of the animals' interlinked hands, touch, and curiosity in their dealings with people and apparently ceaseless engagement with the world. A playful and "even caressing" captive raccoon in Buffon's menagerie was so intent on dipping its hands in water that it was found frozen up to its forearms in ice one cold morning, apparently unharmed. Buffon also quoted at length a letter he received in 1775 from the Frenchman Nicolas Blanquart de Salines, who kept a raccoon as a pet. Salines, much like Linnaeus, claimed that his raccoon's tactile sense dominated all others. "She opens oysters with wonderful dexterity," he wrote. "Her sense of touching must be exquisite. In all her little operations, she seldom uses either her nose or her eye. For example, she makes an oyster pass under her hind-paws; then, without looking at it, she searches with her hands for the weakest part; there she sinks her claws, separates the shells, and leaves not a vestige of the fish." That description matches recent biologists' observations that raccoons, when determining which shellfish to eat, will often decline to even look at their subjects of study, using their forepaws alone to identify species in the muck.[32]

Raccoon hands, then, were not simply a link to their minds: they were also a direct line to their stomachs. Some writers, like Edward Long—a ruthless defender of slavery in Jamaica and occasional natural historical commentator—found humankind's wide-ranging palate to be yet another marker of its superiority over the specialized stomachs of animals, which relied on particular foods to survive. When anticipating a counterargument that man's "stomach confounds him with the other animals," Long insisted that "man alone is unrestrained" in what he can eat, and that "God has given him hands, that he might lay hold of, and fashion, whatever can nourish, cure, and defend him."[33]

Yet the brilliant omnivory of raccoons, key to their success today as well, unseated that argument for humanity's distinction. Raccoons exhibited sensitive knowledge of foodstuffs and their environments, causing them to be celebrated as especially cunning. In 1709, John Lawson found them "very subtle in catching their Prey." In their collaborative *Viviparous Quadrupeds of North America,* the famed artist-naturalist John James Audubon and the minister-naturalist John Bachman explicitly compared raccoons to curious scientists (and to gourmands) when writing that the mussels "that inhabit our fresh water rivers are better known to the Raccoon than to most conchologists, and their flavour is as highly relished by this animal as is that of the best bowl of clam soup by the epicure." Raccoons partly came to be revered as quasi-naturalists due to their proficiency in food environments. For humans and animals alike, a vanishingly thin line separated curiosity, knowledge, and hunger.[34]

Both Audubon and Bachman, however, were slaveowners who also likened raccoon know-how to that of subjugated people and manual laborers. They claimed of the animal bandit: "No negro on a plantation knows with more accuracy when the corn (maize) is juicy and ready for the connoisseur." A shortened version of the word "raccoon" became a derogatory racial slur for African Americans in the nineteenth century; its precise etymology and history is complex, however, having associations with minstrel shows, Whig politics, and possibly (but more tenuously) the Portuguese *barracoon,* a term for slave enclosures. Despite this racialized history, enslaved people *did* have intimate knowledge of raccoon lifeways, which surfaced in or otherwise informed the writings of naturalists. They often hunted raccoons at night, when cover of darkness granted temporarily greater autonomy. A best-selling narrative first printed in 1836 by the pseudonymous Charles Ball, a man enslaved in Maryland, South Carolina, and Georgia who later escaped and published a narrative of his travails, wrote of beginning to "lay and execute plans to procure supplies of such things as were not allowed me by my master," including understanding "various methods of entrapping rackoons," whose flesh provided subsistence and whose furs could be sold. Raccoons, along with other American mammals such as opossums, offered subtle means of material resistance and subterfuge to freedom seekers.[35]

Raccoons also played a more active (if unwitting) role in disrupting the machinery of slavery. Their own knowledge of environments and foodstuffs curtailed the output of plantation agriculture. Along with several other voracious creatures, raccoons relished crops such as sugarcane and corn and domesticated animals like poultry. Farmers today can attest that this is a time-honored raccoon tradition. Across the early modern period and antebellum America—the long lifetime of plantation slavery—overseers sought to manage animal populations alongside enslaved populations. The French traveler Théodore Pavie wrote of a whole conspiracy of animals that destroyed crops on Louisiana plantations:

> Flying squirrels, river rats, and wildcats seem more abundant in wooded places, but blackbirds join them to ravage cornfields. Raccoons uproot the plants before they grow, in order to eat the seeds. In the farthest plantations on the high side of the river, bears devour the seeds. At night when the moon shines, they can be seen sitting down to open stalks of corn with their paws; the planters must send some young Negroes with whips, and the frightened animal flees from the clatter and its echoes.

As Pavie's account makes clear, it was often slaves who were sent as the first line of defense to combat hungry animal populations, especially dangerous ones like bears. Raccoons gained notoriety as some of the prime threats to plantation crops. Thomas Bewick wrote in his *General History of Quadrupeds,* first published in 1790, of the adversarial place of raccoons under the peculiar institution: "The planters consider these animals as their greatest enemies, as they frequently do infinite mischief in one night's excursion." Several prints from the era likewise depict raccoons as gobblers of slave-produced crops like sugarcane (figure 7.8).[36]

When it came to natural history, scientists drew on the knowledge of African-descended people to assess raccoon habits. As intelligent, crafty, and destructive as raccoons were, the animals also took on a tragicomic reputation as fallible. John Lawson, who noted that the raccoon was "the drunkenest Creature living, if he can get any Liquor that is sweet and strong," described raccoons as cunning but also "rather more unlucky than a Monkey." He may have been the first to relay in print an oft-repeated anecdote about raccoons' fateful encounters with oysters fastened to rocks that the animals tried to grab and eat at low tide. He wrote: "Sometimes the Oyster shuts, and holds fast their Paw till the Tide comes in, that they are drown'd, tho' they swim very well." Raccoons came to be seen as tricksters who could be tricked by their own cunning hands. Travelers and naturalists repeated this story for more than

Figure 7.8. From John Church, *A Cabinet of Quadrupeds* (London, 1805), vol. 2. (Biodiversity Heritage Library)

a century, many of them appearing to borrow directly from Lawson and other naturalists' accounts. However, it seems that not all instances of this story were simple rearrangements of Lawson. In 1843, an entry in the *Penny Cyclopaedia of the Society for the Diffusion of Useful Knowledge* attempted to tease out the origins of this popular anecdote of the "living trap." It relayed that the "negroes of Georgia, near Savannah" told a similar tale of raccoon death by oyster, except for those cases when "the four-footed burglar . . . manages to bite off his paw and escape in time." The geologist Charles Lyell, after marveling at raccoon pawprints in sand that led him to oyster beds, said that he, too, heard a similar tale from Black informants in Georgia.[37]

Enslaved people were often derided as manual laborers. But their sensitive knowledge of animals and environments proved vital in assessments of this storied raccoon gaffe—one centered on the animal's own knowing hands. The *Penny Cyclopaedia* continued:

> The planters laugh scornfully at this Negro narrative, treating it as an idle and groundless tale. But zoologists well know that the information obtained from unfettered natives in newly discovered countries, and from those whose condition of life requires them to be much in the open air . . . is not to be despised, and, in the last-mentioned case [of the raccoon drowning], is generally far more valuable than their indolent and luxurious masters can impart . . . [and] with regard to this Raccoon story, the planter deserves to be laughed at more than the Negro.

Whether or not the oyster story should prove to be true, naturalists relied on the intimate animal knowledge held by Indigenous and enslaved peoples when writing animal histories of the New World. People under the bonds of enslavement routinely had more direct contact with animals than did overseers and elite naturalists, as they shooed away bears, hunted animals for subsistence, and cared for livestock. At times, they even had to remind white naturalists to keep studying animals: John Bachman credited his childhood pursuit of natural history to the encouragement of a man named George whom his family enslaved.[38]

Audubon and Bachman, for their part, decided not to take a firm stance on the story of the raccoon and the oyster, having not witnessed it firsthand. But they did otherwise see consistency among raccoon reports. They proclaimed with some fanfare: "From the Alleghany mountains, the swamps of Louisiana, and the marshes of Carolina, we have received nearly the same history of the cunning manœuvres and sly tricks of the Raccoon in procuring food." The raccoon's hand created a world of verification and cross-talk, a world of stories, among naturalists. If any dimension of raccoonhood reached something resembling consensus, it was the animal's sheer tactile guile and ingenuity, its unflagging urge to ask curiously if something might be good to touch and eat.[39]

THE ONES WHO GOT AWAY

Keeping raccoons in captivity or as pets, rather than fleetingly encountering them in the field, emerged as another vehicle for observing these animals and verifying their behavior empirically—or at least for attempting to do so. Even though people certainly still take raccoons into their homes as cohabitants, the custom was remarkably widespread among prominent naturalists and science popularizers in the eighteenth and early nineteenth century. It would be as if Neil deGrasse Tyson, Richard Dawkins, and Sylvia Earle all kept pet raccoons and weren't weird for doing so. Through housing these curious curiosities, historical menageries and homey interiors transformed into domestic laboratories. Here naturalists might test stories about the animals in intimate settings, even if controlled conditions would be a castle in the air.

Naturalists frequently pointed to raccoons' tamability as one of their most salient behavioral truths. Pehr Kalm insisted raccoons were unique in this manner, writing, "Of all the *North American* wild quadrupeds none can be tamed to such a degree as this." De Kay described the animal as "susceptible of domestication," and the terms "tame" and "domesticated" were used interchangeably to describe the raccoon's ability to live alongside humans. According to Audubon and Bachman, the raccoon made a "pleasant monkey-like pet. It is quite dexterous in the use of its fore-feet, and will amble after its master in the manner of a bear, and even follow him into the streets." Much of their joint publication centered on an individual referred to as "our pet Raccoon," though it was likely from Bachman's own menagerie and may have also served as the model for the rotund raccoon profiled in a lithograph accompanying their text (figure 7.9). A raccoon skull uncovered by archaeologists at Jamestown suggests that Anglo-Americans were already keeping raccoons as pets or captives in the early seventeenth century. Native communities throughout the Americas had lived with raccoons much longer, and Marcy Norton has tantalizingly suggested the debt European traditions of pet-keeping might owe to the notion of *iegue,* a Carib term used by various Amerindian groups to describe tame individuals across species lines, ranging from pet-like animals to adopted children.[40]

In Euro-American contexts, people welcomed raccoons into their homes for a variety of reasons, some of them more practical than others. While traversing the American continent in 1802, the French naturalist François André Michaux observed that, taken young, a raccoon "very soon grows tame, and

Figure 7.9. Lithograph of a raccoon, from John James Audubon and John Bachman, *The Viviparous Quadrupeds of North America* (c. 1845). (Courtesy of the Santa Barbara Museum of Natural History)

stays in the house, where it catches mice similar to a cat." Women in particular worked to tame raccoons. Johann Dominikus Schultze, the author of a German-language tract on raccoons published in 1787, said American ladies liked to hold raccoons in place of lap dogs. And William Byrd II wrote with an air of disdain: "Some Old Maids are at the Trouble of breeding them up tame for the pleasure of seeing them play over as many Humorous Tricks as a Munkey." In Europe, raccoons were kept tame in menageries and sometimes placed on display for the eyes of the curious. Captive raccoons might even make a political point. An unfinished sketch from 1797 by Benjamin Henry Latrobe shows the backyard of a Major Watts in Richmond, Virginia. In the yard, an almost apparitional man, possibly enslaved, saws wood; a woman reaches from the house's window to adjust a birdcage; and a pet raccoon sits outside the house in front of its own miniature home (figure 7.10). As Sarah Hand Meacham has argued, keeping pets in the eighteenth-century Chesapeake region served

Figure 7.10. A pencil, pen, and ink sketch by Benjamin Henry Latrobe, 1797. (Courtesy of the Maryland Center for History and Culture, Item 1960.108.1.2.35)

as a means of promoting civility and mastery over both nature and enslaved people.[41]

As ever, raccoon hands appear to have played some role in the initial entrée of many raccoons into human homes. Today the animals are decried as trash pandas in popular culture, liable to conjure visions of refuse and rabies. But in the late eighteenth and early nineteenth century, the raccoon had a reputation as a "very cleanly animal," given its mounting fame for dipping and rolling food and other objects in water with its front paws. The Philadelphia naturalist Charles Willson Peale, for instance, noted of the raccoon in a lecture: "It is very fond of washing its fore paws, will moisten its food by dipping it in water." The American-born naturalist and physician John Godman, who lived a brief thirty-five years before dying in 1830—though not without first leaving a rich record on raccoons based on personal observation—described the animal's "personal neatness and gentle movements," suggesting that not only anatomical similarities in human and animal hands, but also gestures and mannerisms, gave raccoons an air of dainty decorum.[42]

The historian Kathleen M. Brown has shown how Anglophone communities valued bodily cleanliness and keeping the home especially pure. Even as full bathing took a hiatus among English communities in the sixteenth and seventeenth centuries, washing one's hands, particularly before meals, came to be seen as a sign of propriety across the Atlantic world. Some biologists have suggested raccoons may indeed clean some food items such as crayfish. But most agree that raccoons fondle objects in water primarily to enhance the sensory experience of whatever is in their hands, thus deepening their somatic understanding of the world. The overwhelming number of references to the cleanliness of raccoons in natural historical and other accounts suggests an enticing possibility: raccoons may have inadvertently opened a portal into Euro-American homes by enacting their curiosity, performing, with their paws, maneuvers that happened to also accord with prevailing notions of civility and cleanliness at the time.[43]

Many commentators connected the raccoon's use of water to sensation, as when Richard Brookes wrote in 1763: "It makes use of its fore feet in the manner of hands, and feeds itself with them. It have [*sic*] such a sense of feeling, that it will take crumbs of bread, or small bits of flesh out of the water into which they are cast. It is remarkable that it dips all its food into the water, before any of it is eaten." Yet naturalists who took raccoons into their homes or under their care in more public exhibits or menageries began to observe that the "washing" behavior was not universal and perhaps not meant for cleaning food at all. Two raccoons housed in the centuries-old menagerie in the Tower of London "frequently entirely neglected" placing food in water, according to Edward Turner Bennett, despite that behavior being the origin of their Latin species name. Godman noted similar inconsistencies, though he did conclude that water "seems to be essential to their comfort." His own two raccoons, who lived in the yard and were often let loose to run around the parlor, delighted in a tub of water kept full for them, where they could be found "very busily engaged in playing with a piece of broken china, glass, or a small cake of ice." Whereas observations of wild raccoons had created a general consensus of this animal's behavior that became immortalized in its taxonomic name, the keeping of pet raccoons allowed such orthodox beliefs to be challenged through extended exposure to the animals' variable personalities—and, surely in some cases, dispositional changes that resulted from captivity itself. Today raccoons indeed have repute not as conformists but as flexible generalists.[44]

In fact, naturalists frequently noted how raccoons transformed when made tame. Godman observed that "captivity and domestication produce great changes in the habits of this animal, as he learns to spend nearly the whole of the daytime in active exercise, and sleeps during the greater part of the night." Many accounts suggested that not only raccoons' habits but also their disposition altered in captivity. In 1800, the British naturalist George Shaw noted the proclivity of a wild raccoon to stay hidden, especially in winter or bad weather, but he remarked that it became "an active and sprightly animal when taken into a state of domestication," as it was "frequently kept in houses by the Americans." Although wild raccoons were "troublesome," according to Oliver Goldsmith, "in a state of tameness no animal is more harmless or amusing." Raccoons suddenly became playful, mastered tricks, and proved more useful than a house cat.[45]

These passages all indicate humans' faith in their ability to mold nature to their liking through taming. But raccoons' actual *behavior*—evident from reading in between the lines of such testimonies—would suggest humans could not control raccoons as they pleased. One trait naturalists sought to probe and evaluate in a pet-keeping environment was the raccoon's manual curiosity. Naturalists actively introduced stimuli to monitor how raccoons would react. Linnaeus noted that Sjupp would roll a tobacco pipe between his hands for "hours on end," though Sjupp was a "coward" when it came to a hog-bristle brush (leading Linnaeus to suspect some "swine castigated him in America"). Audubon and Bachman similarly recorded that their pet raccoon loved rolling objects in its paws, including cigars, "especially when lighted." Their account gives the impression that any object in the vicinity might be introduced to see—for the sake of human curiosity—what their raccoon's curiosity might do with it. When handed a dead fish, the raccoon acted as an ichthyologist, turning it "in all directions in search of a convenient point of attack. The mouth, nose, fins, vent, &c., were tried. At length an opening was made at the vent, into which a paw was deeply inserted; the intestines were withdrawn and eaten with avidity." When the raccoon was given a jug, "one paw would be inserted in the aperture, and a hundred twists and turns would be made to join its fellow on the outside." They even presented it with a half-grown fox, which made the raccoon "determined to find out the nature of the animal."[46]

But problems caused by captive raccoons' curiosity, intelligence, memory, and sheer resistance and will surfaced in report after report. Recall that curiosity was ever contradictory in the long eighteenth century: a virtue among the learned, but a vice in meddlesome hands. Godman wrote in superlatives: "Nothing can possibly exceed the domesticated raccoon in restless and mischievous curiosity, if suffered to go about the house. Every chink is ransacked, every article of furniture explored, and the neglect of servants to secure closet-doors, is sure to be followed by extensive mischief." Frequent references to servants or enslaved people in accounts of domestic raccoons suggest that well-off naturalists were not the only people coming into conflict with these wily housemates. The raccoon described by Audubon and Bachman remained perpetually chained due to "its well known mischievous propensities." They, like Linnaeus, also noticed the tendency of their raccoon to "adroitly pick its keeper's pockets." Raccoons' cunning and curiosity, but also their volatility and obstinacy, revealed themselves behind closed doors as raccoons dexterously studied their own worlds and the people studying them.[47]

Linnaeus's account of Sjupp paints a picture of a particularly stubborn being. In fact, Linnaeus stated it very clearly: "His *Temperament* was, to be stubborn, and this to the highest degree." Sjupp despised being carried. He held grudges. If one tried to walk him by leash, Sjupp "quickly lied down on the ground, spreading out and resisting" like a latched starfish. When Linnaeus held his late friend Sjupp's remarkable penis bone—which he hailed as "of such a curious sort, that I do not know anything similar" given its bony composition, curved shape, and impressive length in proportion to the animal—he wondered if "he, who seems to have gotten a higher degree of stubbornness, than the other animals, also here should show his temperament" by being, in essence, always at the ready in matters where other animals "must have restraint."[48]

Beyond intractability, naturalists believed raccoons employed subterfuge. Salines insisted his pet, who sometimes liberated herself and terrorized local rooftops for days, used "every artifice to make the fowls grow familiar with her," at which point she would tear the birds to pieces. Raccoons' sneakiness surfaced again in accounts of their thievery. These were not fables, but in-the-flesh *mapachitli*. Kalm wrote that although the raccoon could "in time be made so tame as to run about the streets like a domestic animal," it was unfortunately "impossible to make it leave off its habit of stealing. . . . Sugar and other sweet things must be carefully hidden from it, for if the chests and boxes are not always locked up, it gets into them, eats the sugar, and licks up the treacle [molasses] with its paws." In bourgeois homes, then, pet raccoons gained a reputation as having a sweet tooth, using their hands to secure morsels produced by the labor of slaves on distant yet materially ever present sugar plantations. Kalm reported that ladies, consequently, had "every day some complaint against it, and for this reason many people rather forbear the diversion which this ape-like animal affords."[49]

Raccoon memory posed a particular challenge in the home and menagerie, as suggested by Sjupp's grudges. Sjupp had a long-standing beef with a gardener who once brushed Sjupp off in fear and confusion after the raccoon climbed his leg as a greeting. From that moment on, Sjupp made clear his "relentless hate" toward the man, even if Sjupp so much as smelled him. Shaw concurred that tamed raccoons were "of a capricious disposition, and not easily reconciled when offended." Salines also wrote that his own pet proved "extremely sensible of bad treatment." After being whipped by a servant who then tried to make up for the maltreatment with treats, the raccoon, at the approach of the man, flew "into a rage; her eyes kindle." Such anecdotes reveal not only that raccoons remembered rough handling but that they received such abuses in the first place. Linnaeus's account likewise obliquely mentions that Sjupp received "blows" when the animal misbehaved. Ushering ungovernable raccoons into homes and gardens often created a violent environment for the creatures.[50]

As naturalists came to recognize the willfulness and intelligence of the animals they brought into their domiciles or placed on display, they debated whether raccoons might be permanently tamed or "domesticated" and thus implicitly expressed misgivings that humans could control the species. One observer doubted they would ever be fully retrofitted for human coexistence, noting: "In captivity they become tame and playful, but never domesticated nor particularly attached." Likewise, in his history of the Tower of London's menagerie, Bennett insisted that although raccoons "are easily tamed, and even appear susceptible of some degree of attachment," they

> never entirely lose their sentiment of independence, and are consequently incapable of complete domestication. When placed under a certain degree of restraint they appear contented and happy, are fond of play, and take pleasure in the caresses of their friends, and even of strangers; but however long this kind of domestication may have continued, and how much soever they may seem reconciled to their confinement, the moment the restraint is withdrawn and they feel themselves again at liberty, the love of freedom prevails over every other consideration, and they become as wild as if they had never been reclaimed.

This rights-based language—of liberty and freedom—appears in numerous accounts of captive raccoons and other animals from the pens of British and American authors. It no doubt resonated with a rising movement for the abolition of slavery in the late eighteenth and early nineteenth century. Writers frequently discussed animals in terms of human power relations, invoking enslavement specifically—which, as Ingrid H. Tague has argued, also hastened a movement to treat animals humanely at the end of the eighteenth century. Although raccoons particularly encouraged this rights-based diction given their complex metaphorical and physical ties to slavery, such verbiage spilled over to other animals, especially curious ones confined in homes. William Bartram, a sometimes-slaveowner himself (who called his horse "my faithful slave"), raised a crow named Tom. Bartram praised the bird's "understanding," "cunning," and "wit to select and treasure up in his mind . . . knowledge." Yet Bartram also observed that Tom "aimed to be master of every animal around him, in order to secure his independence and his self-preservation, and for the acquisition and defence of his natural rights."[51]

Bennett's observation that raccoons would inevitably slide back into wildness and that they desired freedom as an end in itself was not purely hypothetical. Ever clever raccoons often escaped captivity. An account in the *Penny Cyclopaedia* recalled "from personal observation" how a free-roaming raccoon, presumably escaped from a menagerie or home, was re-caught in England. Instead of trying to flee again, the raccoon approached the man who discovered him and "ran about his shoulders, put his paws in his pockets, and appeared as much at home as if he had been his companion for years." The author of that publication thus suggested that a once-tame raccoon might remain tame even after liberation. Godman also believed that raccoons did not always run for the hills when given the chance, claiming: "I have had one so tame as to follow a servant about through the house or streets, though entirely at liberty." Godman noted the raccoon was very young when first tamed. It grew so fond of human companionship that it would cry out if left alone. Salines, however—whose pet-keeping narrative came to be excerpted in many natural histories of raccoons (including Godman's text)—contradicted this estimation of the raccoon's permanent tameness. Once again using rights-based language, Salines said that when his raccoon's chain occasionally broke, "liberty rendered her insolent. She took possession of an apartment, would allow none to enter, and it was with some difficulty that she could again be reconciled to bondage." The problem was not that raccoons were untamable. Rather, some simply did not care to be tamed.[52]

One only needs to read a bit further in the account provided by Salines to appreciate the deprivations this raccoon, and no doubt many others, experienced under captivity. Discomfiting as it might be, living with animals forced humans to see their fellow creatures as layered individuals, not as textbook beasts. Caretakers witnessed the full gamut of a creature's life, including the most intimate matters. Salines wrote at length of his raccoon's unruly agitation during mating season. It was a time "when nothing can quiet her." Her apparent isolation from companionship drove the raccoon to masturbation. Salines observed: "A hundred times each day she passes, between her thighs and between her fore-feet, her bushy tail, which she seizes by the end with her teeth, and agitates perpetually, to give friction to the parts." Neither Salines nor Buffon offered additional commentary on this jarring description, which was omitted from most excerpts of Salines published by subsequent naturalists. But pleasuring oneself was yet another hard-to-read trait. On one hand, it was something both humans and nonhumans did, and it implicitly hinted at a sense of self in animals. On the other hand, given masturbation's controversial status in the eighteenth century—as an act condemned by pamphleteers, theologians, and philosophers alike—it too threatened to knock humans down a peg or two in the hierarchy of being by unearthing a baser animal nature. In the mind of Immanuel Kant, one of the leading philosophers of the Enlightenment, it might even cause a person to lose oneself, to commit something worse than suicide, to precipitously fall "beneath the beasts."[53]

It was not only reason, or wisdom, or even cerebral curiosity that naturalists debated as the proper wedge between human and animal—a wedge ever so ready to dislodge. Sometimes, it came back to touch.

Naturalists, driven by their curiosity, restrained the curious and enigmatic raccoon in order to decode it. But it was a tenuous dominion, one that the captives themselves contested in subtle (and not so subtle) ways. Raccoons expressed their individuality and will, which resonated in rights-oriented discourse in the late eighteenth and early nineteenth centuries. Some would acquiesce to domestication, and others would resist, to the point where naturalists literally shackled them in chains to prevent their escape. Even so, many raccoons did break free. Their ability to be tamed would remain a question rather than a fixed and permanent trait to be listed in a book next to anatomical features, diet, and habitat. Raccoons showed pet-keeping to be a permeable prison.[54]

Their ability to exploit the fine line between wildness and tameness has helped raccoons since the Enlightenment era. Unlike rapidly bleaching reefs, fragmented rattlesnake lineages, and declining fish numbers worldwide, raccoons are a rare animal whose global populations are actually blossoming in the face of environmental collapse. They remain in close proximity to people as visible denizens of built environments, thanks in part to the urbanization of America in the antebellum period and the challenges and benefits that human infrastructure present to nimble raccoon hands and heads. Their desire to live on the fringes of human society as commensal shadows, their adaptability, their omnivorous palate, and their curiosity and intelligence have paved a path to success. In some regions, raccoons have so excelled since the early modern period that they have received a badge of scorn as so-called invasive species. Originally restricted to the New World, raccoons now form diasporas in Germany, Spain, France, Russia, Japan, and elsewhere thanks to twentieth-century introductions, accidental and otherwise. Glimpses of raccoon expansion were already evident in the long eighteenth century in the many stories of raccoons in England, Sweden, France, and beyond who, after hitching rides on ships from the Americas, cleverly escaped their European confinement. Some creatures have found ways to shirk the narratives of inevitable decline and shifting baselines that are now commonplace among environmental histories.[55]

Recent research also suggests that today's urban raccoons might be significantly smarter and more curious than their country relatives. Or that is at least the question on the table. Suzanne MacDonald's laboratory at York University in Toronto has found a knowledge "gap" between urban and rural raccoons, the former of which prove far more adept at problem solving with—as should come as no surprise by now—their hands. Challenges to accessing food and the increasing complexity of raccoon-resistant trash receptacles appear to have magnified raccoon intelligence in urban areas. As MacDonald and her co-author Sarah Ritvo write, humans and their cities may unwittingly be "selecting for particular cognitive abilities in raccoons. Persistence, neophilia [love for the new], and high levels of exploratory behavior may result in increased survival and reproduction in the urban setting, and thus we may be observing cognitive evolution in action in this species." MacDonald has called raccoons' hands "their own particular superpower" for exploring the world.

Her pursuit of raccoon minds and malleability can be placed on a continuum with historical investigations of the animals. At the opening of a research talk in 2018, MacDonald told the audience that, in addition to showing slides with graphs of her findings, "I'm basically going to tell you raccoon stories." Much like Linnaeus, Buffon and Salines, Audubon and Bachman, Godman, and the Nahua producers of the Florentine Codex, MacDonald and other researchers must integrate stories of raccoons alongside data to fully encompass their individuality, their ingenuity, and their curiosity.[56]

Buffon believed in an irrevocable gulf between human and animal worlds. This was partly due to the lack of improvement he witnessed among animals. He thought that if animals "were endowed with the power of reflecting, even in the slightest degree, they would be capable of making some progress, and acquire more industry; the present race of beavers would build their houses with more art and solidity than their progenitors; and the bee would daily improve the cell which she inhabits." But he tended to see all animal life, not only that of the Americas, as caught in a downward spiral of entropy. His denial of reason, language, reflection, train of thought, and intellect to animals, and his belief in the divinely ordained dominion of humans over them, however, were not the child of spite; no, Buffon still thought very highly of animals, and he painted a tragic picture of their future in the face of an ever expanding human reign of terror. Buffon believed that, due to human influence, "the talents and faculties of animals, instead of augmenting, are perpetually diminishing. Time fights against them. The more the human species multiplies and improves, the more will the wild animals feel the effects of a terrible and absolute tyrant, who, hardly allowing them an individual existence, deprives them of liberty, of every associating principle, and destroys the very rudiments of their intelligence." But should MacDonald's hypotheses prove true, the intelligence of raccoons—not to mention their curiosity and haptic craftiness—may now be expanding and deepening over time. Perhaps their minds simply haven't changed on the timeline or in the direction Buffon expected in response to human encroachment.[57]

Ideas about animal intelligence and curiosity have an intellectual history. Historians trace it by examining how human conceptions of animal minds change over time. But we can also tell a history of raccoon intellect. Raccoon epistemology, and not merely human epistemology of raccoons, has been subject to forces of historical change.[58]

Buffon, after sketching his vision of the inexorable decline of animals, closed by asking a question for which he offered no answer. "If the human species were annihilated," he wondered to himself and his reader, "to which of the animals would the sceptre of the earth belong?"[59]

Raccoons seem ready to assume the mantle.

CHAPTER EIGHT

Touching the Past

FOLLOWING THE LEAD OF THE CREATURES who have guided us toward the sunset of our journey, I thought I would conclude this whole thing with a success story. Skip the declension and go out with a bang: raccoons on the rise. But I am too disquieted by the changes around me. Not even raccoons, would-be rulers of the world, have so clean an ascension in the Anthropocene, that proposed new epoch in the history of life that bears the mark of the human beast. My efforts to understand the curiosity of modern-day raccoons took me not only to dumpsters around my own home, where plump common raccoons ate their fill. (One such animal led my landlord to post an ominous sign in our apartment entryway that warned: "*BEWARE* Raccoon is out at the Trash Area—Day or Night. *Be Careful.* HE IS EVERYWHERE." She later updated this pronoun in a second poster after the raccoon gave birth in the trash bin.) Not content merely with *Procyon lotor,* capacious as that creature is, I also had to journey much further afield to the island of Cozumel to get a better grasp on the relationship between raccoon knowledge and scientific knowledge today.[1]

Cozumel is a limestone oasis located off the coast of Mexico's Yucatán Peninsula, the easternmost edge of the country that points into the Caribbean Sea like an upturned boot. Measuring thirty miles long and ten across, Cozumel, like many islands, hosts a number of endemic species, animals found nowhere else in the world. And one is a superbly special raccoon. *Procyon pygmaeus*—known variously in English as the Cozumel raccoon, pygmy raccoon, and dwarf raccoon—has become an unofficial mascot for the tropical isle. I saw its bandit face stare back from t-shirts and signs all throughout the coral-flanked landmass. Like many island animals, these raccoons are significantly smaller than their mainland cousins, the common raccoon and the crab-eating raccoon. They are also morphologically distinct in more compelling ways, as Cozumel raccoons sport remarkable golden tails. And, like many an island vertebrate of the twenty-first century, they aren't doing too well. As few as two hundred mature pygmy raccoons may cling to existence, far and away making them the world's rarest raccoon species and earning them a designation in bright crimson as Critically Endangered on the Red List of the International Union for Conservation of Nature, or IUCN. Nationally, Mexico's environment ministry (La Secretaría del Medio Ambiente y Recursos Naturales, or SEMARNAT) classifies Cozumel's raccoon as "En Peligro de Extinción," in danger of extinction. By contrast, the common raccoon not only escapes the perilous red portion of the IUCN's list but is described by the organization as *increasing,* signaled by a heavenward green arrow.[2]

As with our research foray in tropical Australia, we would first have to conquer a vomit-inducing boat and treacherous dirt road to meet the Cozumel raccoon. William and I were visiting my father-in-law, Guillermo, in Mexico at the time. To get to the island, the three of us boarded a ferry from Playa del Carmen in the state of Quintana Roo, though one can also travel to Cozumel by air or by cruise. A Maya spiritual site for thousands of years with a significant ethnic Maya population today, Cozumel also ranks as one of the most important cruise ship docking sites globally. Millions of vacationers per year disembark from these floating hotels, whose titan anchors crush the bones of the island's famed corals below. Our short naval passage was almost as rough as the waves we encountered near Endeavour Reef. Crewmembers preemptively distributed barf bags. Many travelers vigorously indulged. As we left port, sea-legged workers in Cozumel's tourist trade started wandering the floor of the vessel, offering local activities like dolphin canoodling to green-faced passengers. For each sales pitch we received, Guillermo would return a seeming non sequitur: "¿Dónde están los mapaches?" Finally, we found Carlos, a Cozumel resident who approached us hawking other experiences. He said he knew the raccoons well.[3]

Once we disembarked, Carlos led us to his car, a low-riding sedan that was most certainly not an SUV. We started on a bumpy, half-flooded, crater-bedecked path through thick trees and thick air. In time, a dock materialized with a few weathered fishing boats but no wildlife in sight, save for five vultures suspended above us. Some teal plastic chairs waited emptily near the shore. A deflated soccer ball and discarded cups littered the ground.

I assumed the chance of seeing these critically endangered raccoons would be slim anywhere, especially in this lonely port. (Taking a hint from rattlesnake protectors, I've kept the details of this location vague enough, though locals, I'm sure, may know the port of which I speak.) Prior web research suggested that perhaps some of these raccoons could be found near the island's tourist hubs, restaurants, and garbage cans—although this suggestion provoked my skepticism too. The rarest of rare species are supposed to be shrouded in a cloud of mystique: a remote camera trap triggered by a ghostly Amur leopard in Russia's Far East, or the wing rustle and peripheral glimpse of an ivory-billed woodpecker, a fowl so elusive that its extinction has been debated for decades, leaving written accounts of the bird to continually waver between "was" and "is." If we are to see critically endangered animals at all, we expect to see them in zoos, the magic of encounter drained by the ease of our viewing and the distance of bars or safety glass. But all Carlos had to do was crinkle a plastic bag he found on the ground as a summons. Moments later, two raccoons in miniature shot out of the mangroves and toward us like terriers. As did another. And then another. And then another. Oh me of little faith—who was I to doubt Carlos, who had visited this spot so many times before us, the place where he took his children to meet with *los mapaches enanos*?

One, two, three, four, five, six, seven, eight, nine. Whatever the real population numbers of the species, I could count nine in this remote island corner—around 5 percent of the global adult population tallied on my own two hands, if the bleakest estimate of the remaining survivors is correct. The one with the graying face. The one with clumps of fur missing from its back. The one who loved fruit. The one who lazed on a rock at my knees (figure 8.1 and plate 13). The two who insisted on rolling their food in the sheltered water beneath the mangrove trees, their core habitat, like the washing bear of Linnaeus. The one who snarled at the others. The nine with the golden tails.

A large wooden sign warned us not to feed these endangered creatures. One of its rare referents scurried beneath the frame. Such assistance, the sign claimed, would incapacitate the raccoons for *la vida silvestre*—a phrase usually meaning "wildlife" but in this sentence signifying "life in the wild"—and thus jeopardize the survival of the species as well as Cozumel pride, or *nuestro*

Figure 8.1. The Cozumel raccoon (*Procyon pygmaeus*) who loved the rock. (Photograph by Whitney Barlow Robles)

orgullo cozumeleño. A *National Geographic* guide to ethical wildlife photography written by the journalist Melissa Groo similarly cautions: "The kindest thing we can do for wild animals is to honor their wildness. The quickest way to compromise that wildness is to offer food." However, this guide also advises honesty with one's audience: disclose the conditions of possibility for your animal encounter. If you fed them, let us know.[4]

When I once tossed some bread to ducks in a Massachusetts pond, I became the target of a drive-by scolding as a harried Englishman on a bicycle whizzed behind me and blurted a staccato "please don't feed the ducks" faster than I could turn around to make eye contact. I had so often received the mainstream environmentalist message that one should not feed wild animals, which would render them helplessly dependent on or dangerously habituated to people—the idea that, once it takes food from human hands, an animal ceases to be wild, or worse: it ceases to be natural. Food has certainly brought humans and other animals into potentially violent familiarity, as shown by campaigns to stop people from feeding bears in national parks or, at the very least, to secure trash cans and retire bird feeders in bear country. It's not that well-fed bears compulsively maul anyone who steps in their path. But the newly visible bears are deemed too comfortable with our species, too comfortable approaching people or human habitats for food, a behavior that often prompts euthanasia by authorities. A fed bear is a dead bear, so they say.

Prohibitions around feeding animals reveal as much about our conceptions of what is "wild" and what is "natural" as anything else. Some are yet another instantiation of the wilderness conceit deconstructed by William Cronon, whose creaturely counterpart would be a theoretical pristine beast, untainted by human influence. The London-based Wellcome Trust recently funded an initiative to probe these questions, showing animal feeding to be a fertile new research area. Seated at the University of Exeter and other U.K. institutions and titled "From Feed the Birds to Do Not Feed the Animals," the project seeks to unravel why humans seem predisposed to feed other animals. Combining historical, sociological, anthropological, and scientific perspectives, the project's research highlights when animal feeding has been beckoned and when it has been banned and considers what the environmental or health costs of feeding domesticated animals, birds, captive zoo life, and wildlife might be. The researchers center the essential contradictions that swirl around animal feeding, as evident—to give one of their examples—in zoos that post signs warning "Don't Feed the Animals" while they sell buckets of food at the front desk for visitors to do just that. To their point on the endurance of this theme: my own father lovingly makes scrambled eggs for the raccoons who visit his back porch every night and handfeeds them peanuts—activities that saw an uptick during his extensive pandemic isolation. "The kids like fries and tater tots," he told me, "but not mom. They all like peanuts, though."[5]

Much like the sign we encountered in Cozumel, a tweet from the Parque Nacional de Arrecifes de Cozumel in 2015 condemned feeding the island's raccoons, itself evidence that many people continue to provide them succor. The tweet branded the act a potential extinction vector, citing health impacts from the incompatibility of processed food with a raccoon's omnivorous diet. (As would befit the inherent contradictions of animal feeding, gastrointestinal tracts of dumpster-diving common raccoons rarely merit the same compassionate concern, despite the fact that the discarded junk food and sugar of our species wreak similar havoc on their bodies.) Cozumel's park authority also condemned human feedings for their potential to make the island's raccoons lose not just their hunting ability but also their predation-protective *miedo al humano:* their fear of humankind.

These cautions, of course, all have truth in them. But they also assume that fear of humans is the natural order. Raccoons and people have been negotiating their fears of one another and overcoming them to coexist for many centuries in the Americas, sharing food, homes, and embodied touch. In some cases, they have devised a new natural order of mutual companionship. If you like, call it love. There is an idealized Cozumel raccoon. And then there is the one that must make do in the world it actually inhabits. With other human actions so severely imperiling Cozumel raccoons, is it really any wonder these animals would look to people for a little help, given that raccoons have been breaking into human foodstuff for centuries? Might we not owe them a debt, in the short term, in lieu of or alongside structural changes to rebuild their populations in the long term? And would it not befit their nature, as some of the craftiest American animals, to procure payment of this debt with an open palm?

Carlos had preemptively grabbed twenty pesos' worth of fresh tortillas from a local *tortilleria* on our way to the port. He always did so on days when he took his family to see the raccoons. I had already snuck several from the warm bundle on our drive. Carlos knew that these raccoons, like those that raided eighteenth-century plantations, loved maize, even though a typical Cozumel raccoon's "wild" diet consists of mangrove crabs, insects, an array of fruits, and the occasional sea turtle egg. We had fruit in tow as well. Perhaps the whole scenario would have been a different matter if these raccoons weren't so insistent, so stubborn like Sjupp. But when they approached us, restraint was futile. With warnings in plain view, I tuned out my better angels and did what so many people had done before me.[6]

I fed the multitude. Only my fish and loaves were fruit and corn masa.

To my knowledge, these raccoons were the rarest animals I had ever seen, even scarcer than an Australian cassowary. They were also, paradoxically, the tamest. The first thing you notice is how much they touch you. Their paws are soft and padded, like the worn-in leather of motorcycle gloves. They place their hands directly into yours, palm on palm, forging mutual recognition through holding more than beholding. Knowing a bit about raccoon biology and the history behind these wandering hands, I interpreted such gestures as their way of locating food and of getting to know who or what I was (figure 8.2 and

Figure 8.2. The author communes with a Cozumel raccoon. (Photograph by William Robles)

plate 14). They certainly demonstrated that neophilia, that love of the new, that insatiable curiosity discussed by Suzanne MacDonald as a characteristic raccoon trait—one that sets them apart from most wild animals. Sometimes standing erect to amble about on two legs, golden-tailed raccoons are walking contradictions.

The longer we stayed, the more comfortable they grew. There was something tender and intimate about these encounters, standing hand in hand with a raccoon on the brink of extinction. Several rose and rested their palms on my knee, using my leg as a brace in hopes of fruit. If I made a tortilla-filled fist, they would pry my fingers open like a clam. Occasionally one would approach, tap its hands into my own and, sensing they were empty—and perhaps also sensing I was no threat—waddle casually on. More than once, my finger slipped into a mouth. (I probably now have some rare disease.) Another standout feature of this isolated dwarf species, in comparison to other raccoons, is what biologists call their "remarkably reduced" teeth. The entire time we stayed with the raccoons, we were connected in fleeting but firm moments of touching.[7]

Indeed, I had to keep figuratively pinching myself, bringing to the fore of my mind with some conscious effort amid repeated endangered raccoon pats that this was a deeply bizarre situation to be in. It only really seemed possible because they were raccoons, uniquely capable of combining extreme rarity with extreme familiarity. These animals have always been getting into human food. It's almost what a raccoon is. They did so with Native, European, and settler communities centuries ago—food likewise serving there to bind humans and raccoons and fuel each party's curiosity of the other. I slaked my own thirst for knowledge of these animals through shared sustenance and shared touch. The appetite of the body and the mind collapse into one.

Biologists who are currently working to protect these animals today analyze their diets alongside samples of their blood, tissues, and fur. They parse Cozumel raccoon stomach and fecal contents with gusto. How do they obtain samples? Well, they use food—sardines, but also banana and pineapple, fruits like we had offered—to bait Cozumel raccoons into live traps for study, genetic analysis, and scat collection. Yet this form of animal feeding is sanctioned as research. Politics of authority determine who gets to feed animals permissibly, using "appropriate" foods, and to what end. The same biologists who feed these raccoons, for instance, insist human-raccoon interactions on the island must be managed and limited, given that raccoon populations typically become much denser—and notably more family oriented—in the presence of humans than in their absence. In the case of the Cozumel raccoon, biologists suggest this trend could open the door for diseases to course through the community or worsen genetic isolation. "Therefore," they conclude, "anthropogenic food sources and human interactions . . . should be significantly controlled and reduced." Indeed, locals like Carlos are faulted for their family tradition of breaking bread with these raccoon families, even as Carlos was my guide and gatekeeper for accessing such endangered animals in the first place.[8]

Is this, then, what extinction looks like? Not only bleached reefs and secret rattlesnakes, but precarious fauna popping jack-in-the-box style out of garbage

Figure 8.3. Two Cozumel raccoons approach the author. (Photograph by Whitney Barlow Robles)

cans? Such critical endangerment, yet such brazen boldness, all wrapped in one soul (figure 8.3)? Despite recent pleas to separate people and Cozumel raccoons, the answer seems to be a tentative yes: we have seen that to be a raccoon, centuries ago as much as now, is very often to be embroiled in human affairs. As with the corals I encountered in Australia, here I felt a mixture of hope and loss when faced with these beings teetering on the edge of being. I was aware of their singular capacity to adapt to humans and development. Unlike other critically endangered animals that loom in the shadows, even the most endangered raccoon has found a home among people. Perhaps they will forge a new way, one hand in hand with our species, as tourism rapidly alters the island. Raccoons don't fit calmly into archetypes of wild and tame, natural and artificial, because they are ever liable to transgress those boundaries and dissolve them altogether. As the Nahua authors of the Florentine Codex said of the raccoon, or *mapachitli,* in sixteenth-century Mexico City: "In forests, crags, the water, among reeds, everywhere it lives. Wherever it feels comfortable, there it lives. . . . And since it is a great thief, since it takes everything that it sees, and since it has human hands, a thief is also called *mapachitli.*" Read in this light, raccoons stealing handouts from us seems most true. If raccoons become a little more human and humans become a little more raccoon, there might be a road to coexistence.

Lulu Miller's short story "Animal Planet" unknowingly takes up Buffon's final inquiry by contemplating how animals might feel were humankind to go the way of the dinosaurs. In this vision of our own extinction, owls rather than raccoons rule the earth. But the bandits still have a role to play. Given the imbrication of touching and eating and knowing and living, raccoons are the sole animals to grieve humanity's absence. "They had built up mythologies about us as The Benevolent Giver," writes the unnamed narrator. "Now, they convene in our abandoned churches and pews. While the rest of the animals rejoice as our rotting flesh fertilizes the soil, the raccoons alone pray for our resurrection."[9]

Although humanity's last vanishing act is a useful thought experiment, the imminent slippage of other animal species from existence is sadder than fiction. On Cozumel, mammals like raccoons, along with the even rarer dwarf coatis, are especially at risk. New development of hotels and docks and tourist sites carves up their habitat and fragments their populations into small genetic islands, perhaps hastened by the feeding of people like myself. An expanding road system results in fatal vehicle collisions for the animals. Climate change and its worsening hurricanes materially harm the raccoons and erode their habitat. Human-introduced predators like boa constrictors and feral dogs devour them. So-called congeners, in the form of wayward mainland raccoons who travel to the island as pets or by other means, threaten to put into question even what a Cozumel raccoon is, due to hybridization and genetic introgression caused by interbreeding.[10]

Is it extinction or survival if they shapeshift into something else?

In the face of these and other threats, nine raccoons, or even several hundred, may not be enough to withstand the challenges ahead. Their existence on paper is currently up for debate, mirroring the shifting nomenclatural status of raccoons in the eighteenth century. In 2020, a paper in the *Journal of Zoological Systematics and Evolutionary Research* questioned whether the Cozumel raccoon should be considered a separate species from the common raccoon at all. The stakes are high: welcome these golden-tailed wonders into the fold of *Procyon lotor,* and they will melt off the IUCN's Red List and forgo major protections. They'll become just another pest. In 2022, another team of ecologists and geneticists swiftly responded with a rebuttal paper and larger sample size arguing for the Cozumel raccoon's distinction.[11]

Wherever the taxonomic cards fall, I have an earnest wish for the kin of the nine that I met. Let us hope that these earthly things will not recede from the tree of life like waters rushing out from beneath the mangroves they call home.

Epilogue

SHE MIGHT HAVE BEEN BORN during Columbus's final voyage. Nearly blind from parasites, she mated with what scientists not unromantically call a "distinct pairing with embrace." She bore children that were still considered children at more than a century old. Her flesh was highly toxic. She reached her enormous length at the unhurried rate of one centimeter per year. People have opened the guts of her kin to find the jawbone of a polar bear, whole reindeer in silent repose, even a reported human leg. She was and did many other things we don't know, don't understand. Sometime between 2010 and 2013, she died. This we do know, from the backmatter of a paper published in 2016 in the journal *Science*. The study found her kind, the slow-moving Greenland sharks of the Arctic Seas, to be the longest living of all vertebrates (figure E.1). Described formally as *Somniosus microcephalus,* they are also some of the largest sharks in existence, capable of surpassing twenty feet in length. That fact, paired with their protracted rate of growth, tipped off scientists that such animals might be ancient indeed.[1]

How this shark fell into the hands of researchers is not apparent from the main attraction. One must locate the article's less-trafficked Supplementary Materials, which note that she and the twenty-seven other female Greenland sharks assessed by the study were acquired as "unintended bycatch during the Annual Fish Survey of Greenland Institute of Natural Resources, by the commercial fishing fleet and from scientific long lines." Researchers, drawing on

Figure E.1. A Greenland shark, with a parasitic copepod attached to its cornea. (Photograph by Wikipedia user Hemming1952, via a CC BY-SA 4.0 license)

the networks of foodways, euthanized those of the mistakenly captured marine methuselahs that had injuries presumed to be lethal. Then they removed their eyes. Based on an innovative analysis of radiocarbon isotopes in ocular tissue, this particular female was estimated to be 392 ± 120 years old: 272 years old at the lowest estimate and 512, a regal half-millennium, at the highest. That means this individual shark was swimming around during most of the historical events explored in this book, submerged leagues beneath the feet of humans on shore. Perhaps she bisected the shadows of trade ships and slave ships ferrying naturalists across the Atlantic. She was spectator to Enlightenment natural history, just as others of her kind, like the 13-foot-long Greenland shark laconically described in the *Systema Ichthyologiae* of Marcus Elieser Bloch and Johann Gottlob Schneider, were also its barely figured participants. In this case, the live historical animals prove more resilient than the preserved ones: the dried holotype specimen of the species from Bloch's collection was last documented in 1841. Despite its impressive size, it has since gone missing, likely discarded by nineteenth-century curators for its unsightliness.[2]

The lifespan of a mollusk known as the ocean quahog (*Arctica islandica*) also tops the half-millennium mark, besting even the Greenland shark. And a single "colony" of black coral from dim depths near Hawai'i has been dated at more than four thousand years old, confounding our definitions of both longevity and bounded individuals. You may also recall that a polyp, under the right conditions, could theoretically live forever. In figurative but also literal ways the animals of the Enlightenment are still with us. We are enmeshed in their history, they in ours. These selfsame animals were present during a history that seems remote. Then again, given that centuries-old sharks are today being caught and killed in nets as bycatch, despite their categorization as "Vulnerable" on the IUCN's Red List, while deep-sea corals face existential threats from bottom trawling and the commercial jewelry trade, we can see how these enduring figures are not timeless or static or renewable but also threatened and transformed by changing historical circumstances. They have, and make, and tell history. Indeed, one crucial factor in that shark study's radiocarbon analysis was the so-called bomb pulse, a swift increase in the radioactive isotope carbon-14 from nuclear testing and nuclear fallout in the middle of the twentieth century that can be read like a specter in tissue. Humankind's ultimate capacity for destruction is yet another historic event witnessed by these centuries-old sharks and archived in their bodies. The radical difference of animals—their disparate life cycles, their divergent sensoriums, their distinct experiences of time and space—asks us to think differently about our definition of history, and about when an era has passed.[3]

From the reefs of the Pacific, to the cabinets of Europe, to the caverns, kitchens, and cages of the Americas, animals snared humans in a web of influence across the long eighteenth century. Animals collected discoveries, shared homes with naturalists, inspired colonial visions, and quite literally electrified the search for knowledge. They also crippled ships, ate priceless specimens, stymied the labors of printers and collection keepers, made the throats of men tighten with fear or disgust, and arrested the human search for knowledge in other visceral ways. We've seen the stories of compelling animal individuals like Sjupp and I.II.I.a, along with anonymous coral polyps and long-dead fish skins, each with its own vitality, its own mark left on some corner of Enlightenment-era natural history. And we've seen a menagerie of human interlopers engaged in the study of nature: not only the rarefied Linnaeus and Buffon, and not even just those lesser-known Royal Society dabblers and colonial travelers. We've also met Blanket, whose rattlesnake knowledge preempted modern herpetologists by several centuries, and anonymous chefs and unfree collectors whose fish creations live on, not merely cut-and-dried, in Linnaeus's specimen vault.

This book has studied scientific knowledge as a more-than-human endeavor, the messy lovechild of human culture and animal power. Curiosity and animal nature mingled to give birth to natural history. Animals added considerable friction to the quest for knowledge in the early modern world, just as often presaging its loss as its generation, whether that knowledge served to enable imperial expansion and dispossession, to place people in hierarchies, to advance a program of nationalist science, or as an end in itself to write (or find oneself unable to write) the natural history of a particular creature. And this history of knowledge, of knowledge as the mixed brood of curious species, bleeds into the contemporary. We have encountered the descendants of these historical animals, some emboldened and some worse for wear, some rearing their heads in uncannily familiar ways, like the abandoned lumpfish in Joe's freezer. We've heard tales from the crypt, uncovering long-lost specimens while puzzling at the absence of others. We've seen multiple lines of expertise and a diverse array of human informants not unlike what we witnessed of the eighteenth century: state-sanctioned scientists, those brandishing PhDs, and those with university posts, but also animal experts who are local, Indigenous, commercial, or lay. And we've seen the personal, physical, and emotional dimensions of natural history across time. Animals, as much as we'd like to study them for their own sake, also hold a mirror up to ourselves.

Can a brainless polyp choose? Do rattlesnakes enchant? Might dead flesh beguile? What is a raccoon, and what might it know? Eighteenth-century animals created an impasse about the capacities of the nonhuman world that has never fully been resolved. The unknowability of animals persisted into the nineteenth century, and at the beginning of the twentieth such questions escalated into the so-called nature fakers controversy. This explosive moment in American literary and cultural history served as a public referendum on whether animals could pull off seemingly anthropomorphized feats, such as a porcupine rolling downhill for no other motivation than exhilaration, or a woodcock making a cast of mud and reeds for its broken leg. Of all places, the detractors of these sensational animals came from the genre of nature writing. The famed naturalist and essayist John Burroughs led the charge in March 1903 with an *Atlantic Monthly* diatribe titled "Real and Sham Natural History." Burroughs identified popular writings about partridges taking roll call, foxes riding on the backs of sheep to confound hunters, and crows holding court as "mock natural histor[ies]" meant to willfully deceive the public. Nature writers and scientists clamored to the press to debate his claims. In 1907, President Theodore Roosevelt entered the fray by announcing his support of Burroughs. The years-long brawl over animal powers left Burroughs breaking down in tears in public, and William J. Long, one of the controversy's primary targets, reportedly went blind from the stress of the ordeal.[4]

Today's historians have reached no consensus on animal agency. Scientists, too, continue to debate the capabilities of animals. In his book *The Evolution of Beauty* (2017), the Yale ornithologist Richard O. Prum argues that birds' aesthetic choices play a central role in the formation of species. Prum believes that "animals are agents in their own evolution." But forays into the strange wonders of animal thought have been slow and tepid. In Prum's estimation, that's because these experiences are unmeasurable within the normal parameters of research and therefore most scientists remain "allergic to the idea of making a scientific study of subjective experiences, or even to admitting that they exist." Titles from the last few years such as Jonathan Balcombe's *What a Fish Knows: The Inner Lives of Our Underwater Cousins* and Frans de Waal's *Are We Smart Enough to Know How Smart Animals Are?* (both published in 2017) reveal the continued salience of animal power as a question of epistemology. It's as if we're always catching up to the fact that we're always catching up to animals.[5]

And these examples are only about animal minds. There remains an even deeper, more intractable problem of nonhuman matter and mattering, of material resistance and alter-agency that may have no direct link to conscious thought but which is also our legacy from the animals of centuries past. Age-old anxieties about nature's sway over humanity plague us still. Invisible viruses, which stalk the border between the living and non-living, continue to expose our animal nature and inability to control or comprehend the natural world. Our bodies are not fully our own, teeming with lifeforms that infiltrate our behavior, psyche, and ability to know things, as recent microbiome research suggests the quiet role of gut bacteria in conditions as varied as Alzheimer's, autism, and Parkinson's. We are every day navigating the currents of nonhuman influence and the animal foundations of our knowledge and the prying open of what it means to be human that unsettled many a thinker of the eighteenth century.[6]

Yet for all the power of animals, one need only look at the accelerating destruction of coral reefs worldwide, or the plight of timber rattlesnakes, or the collapse of fisheries, or the handful of Cozumel raccoons left scurrying around the island to conceive of animals and their communities as finite historical entities that change over time. Their stories show the resilience of such creatures in the face of radical change and suggest how the fates of animals are bound with our own. *Curious Species* has explored the curious animals and curious humans of the past, in the hope we might fathom curiosity's promises, edges, and limitations for an uncertain and ever warming future.

Let's conclude with a portrait of one final animal. His forepaws grip tightly around an object of study, his precious relic of focus, which he has alternately smelled, surveyed, touched, and tasted. His hair is wild and coarse. He is hungry. He is tired. Perhaps he is fearful, too. His animal body wearies from exposure to the camphor and mercuric chloride protecting his midden, his cabinet. His habitat is various, terraqueous. Habits mostly diurnal. Diet omnivorous, with some exceptions. To quote Linnaeus, who gave the species its still-recognized name: "*Fore-teeth cutting; upper 4, parallel; teats 2 pectoral.*" The Swede wrote additionally of this European variety's protective outer coating of "close vestments," of its fair skin and blue eyes and flowing hair, of its distinction as being governed by laws or rites instead of the "customs," "opinions," or "caprice" that supposedly rule other varieties. Prejudice is also an ingrained trait in the species. This particular specimen's snout is prominent, eyes close-set, skin wrinkled and wizened. Myth and legend haunt his natural history as well as observable fact; some say that he will be killed for being a god, others that he fancied himself a demigod. We're still not sure. There are other things we don't know of his nature.[7]

This portrait is a composite. Its qualities could reflect any number of the naturalists in this book: John Ellis's multisensory coral study, or Benjamin Smith Barton's snake phobia, or Joseph Breintnall's nightmares, or William Dandridge Peck's occupational hazards, or the fish fondling of either Gronovius, or Captain Cook's deadly fate in Hawai'i, or even Linnaeus's famed arrogance. Some scientists suggest Linnaeus could be considered the lectotype specimen of *Homo sapiens*—a retroactive designation when no originary holotype specimen has been specified. As an animal himself, this amalgam naturalist fell under the project of natural history too. If he sits, in his mind's eye, uncomfortably at the top of the Chain of Being's animals, it is a cage of misery in which he has chained himself while the rest of the animal kingdom flies unbridled and scurries amok and swirls in murmurations around him. The greater part don't think of him. He is altogether curious. He is not alone on this earth.

ACKNOWLEDGMENTS

Like some of the naturalists we encountered in the preceding pages, I cannot help but compare a monumental project—in this case, the writing of a book—to the animal labor of a coral reef: a cumulative process that takes time, community, grit, and a little bit of salt. Many small contributions have, bit by bit, assembled something massive. If I'm lucky, it will be strong and long-lasting, too.

First and foremost, I must thank the dozens of individuals who made this reef possible, and especially my family. My husband, William Robles, offered feedback on many parts of this manuscript and helped me trudge through arguments as they came into being. He was my right hand on numerous trips to archives and field sites. He followed me to abandoned car parks stashed with hidden paintings in Den Haag, took expert photographs of documents with me throughout London archives, and drove us safely through (yes, drove, yes, through) crocodile-infested waters in Australia, among so many other labors of love. I am eternally and happily indebted to his devotion and fandom. My parents, Kristine Nagus and Michael Barlow, my brother, Evan Barlow, and my in-laws, Winora, Guillermo, and Janell Robles, offered unconditional affection and care. Our feline family member, Dora Robles, sat by my side (and often on my computer or handwritten outlines) for much of this book's composition. She has been my working companion; a surprisingly versatile conversationalist when I needed a sounding board amid the fog of writer's block; at times, the fool; and an ever-present model of the rich inner worlds of nonhuman animals. I would not have been able to finish the book without loving childcare from Melissa Robison, Katy Tremblay, and Ethan Sullivan-Dupuis. My daughter, Luna, has only been alive for the tail end of the writing process, but she has left her tiny marks all over this book as well.

I have benefited from incredible communities of human animals at Harvard University, Dartmouth College, and beyond. I am particularly grateful to the interlocutors at my book manuscript workshop—Daniela Bleichmar and Bathsheba Demuth at the helm, along with Leslie Butler, Mackenzie Cooley, Danielle Simon, Rebecca Clark, and Amy Schiller—for helping to shape a late-stage draft into what this book is now. In addition, Joyce Chaplin, Harriet Ritvo, Jennifer Roberts, Janet Browne, Irus Braverman, Grace Kim-Butler, Carla Cevasco, Zachary Nowak, Catie Peters, Maddie Williams, Chloe Chapin, Alicia DeMaio, Pete Pellizzari, John Bell, Megan Black, Emilie Connolly, Liz Polcha, and Diana Epelbaum all provided extended commentary on drafts. For their critically useful feedback in the form of comments on my writing, questions during talks, or suggestions for further research, I thank Sheila Jasanoff, Jill Lepore, Anya Zilberstein, Hannah Marcus, Paul Musselwhite, Charlotte Bacon, Colin Calloway, Thomas Wickman, Strother Roberts, Virginia Anderson, Jane Kamensky, Maurice Crandall, Laura Ogden, Warwick Anderson, James Delbourgo, Christopher Parsons, Deirdre Cooper Owens, Dolly Jørgensen, Dan McKanan, Alex Csiszar, Darrin McMahon, Marcy Norton, Molly Warsh, Mayra Rivera Rivera, Susan Scott Parrish, Laurel Thatcher Ulrich, Sara Schechner, Ian J. Miller, Ann Blair, Dániel Margócsy, David Hackett Fischer, Zeb Tortorici, David Jones, Tiya Miles, Jim Hanken, Deidre Lynch, John Dupré, Mary Terrall, Stefan Helmreich, Jim Warren, Steve Swayne, Greta LaFleur, Kathryn Braund, Randy Harelson, Simon Sun, Deirdre Moore, Nicholas Rinehart, Bruce Duthu, Preston McBride, Jeremy Mikecz, Hannah Anderson, Camden Elliott, Amy Fish, Luke Willert, Christopher Allison, Jordan Howell, Shireen Hamza, Elaine Ayers, Kenneth Cohen, Chris Rodelo, Lucie Steinberg, Laura Nelson, Jenesis Fonseca, Andrew Block, Eva Payne, Allison Puglisi, Edwin Rose, Ardeta Gjikola, Iman Darwish, Gustave Lester, Ben Silverstein, Marion Menzin, and Chris Baker. Jens Amborg translated dense eighteenth-century Swedish for me. And I am indebted to the support of many wonderful administrators, including Arthur Patton-Hock, Monnikue McCall, and Laura McDaniel.

I must also give special thanks to Linda Peterson. She advised my undergraduate thesis and tragically lost her battle to cancer near the beginning of my graduate career. She introduced me to the genre of nature writing when I took her undergraduate literature course at Yale University, and her mentorship helped put me on a path to pursuing a PhD. I learned of her death from a co-panelist only minutes before I stood up to give my first conference presentation as a graduate student. She is dearly missed.

Portions of this project have appeared in previous publications, and comments I received throughout the editing and review process at each greatly strengthened the book. I especially thank Joshua Piker, Meg Musselwhite, and seven peer reviewers for the *William and Mary Quarterly*, including David Gary Shaw; Ellery Foutch and Sarah Anne Carter at *Commonplace: The Journal of Early American Life;* and reviewers of a book chapter that appeared in *The Philosophy Chamber: Art and Science in Harvard's Teaching Cabinet, 1766–1820*, published by the Harvard Art Museums and Yale University Press. I thank all of these publications for permission to reprint portions of those essays here. Feedback I received through conference presentations and other talks has been invaluable, too. I am grateful to audience members and fellow panelists at meetings of the Museum of Comparative Zoology Seminar Series at Harvard, the American Historical Association, the American Studies Association, the Omohundro Institute of Early American History and Culture, the McNeil Center for Early American Studies, the History of Science Society, the Association for the Study of Literature and Environment, the Joint Atlantic Seminar for the History of Biology, the Colonial Society of Massachusetts, the Water History Conference, Harvard University's Early Sciences Working Group, the Program in Native American Studies at Dartmouth College, Harvard University's Early America Workshop, Harvard University's STS Circle, and the Bartram Trail Conference.

It has been a joy to work with Yale University Press. I thank my editor, Jean Thomson Black, for being an early champion of the project; the expert eye of editorial assistant Elizabeth Sylvia, who fielded countless questions about formatting and image permissions; and Liz Casey and Phillip King for their careful attention to the manuscript. I thank everyone at the press for their enthusiasm for the book, and especially the external reviewers, including Cameron Strang, for investing time and energy in it.

People beyond academia have shaped the book as well. My good friend Zachary Rotholz got his hands dirty—literally—to assist me in reenacting historical taxidermy procedures. His boundless creativity is something we should all aspire to. Inspiration for this book also came from my earliest exposures to science writing and editing: at Caltech with Rustem Ismagilov and at the American Museum of Natural History with Lauri Halderman, Jane Levenson, JoAnn Gutin, Joan Bernard, Anne Canty, Sasha Nemecek, Martin Schwabacher, Vidya Santosh, Eleanor Sterling, and Stephen Quinn. They have left a stronger mark on this book than they probably realize. Lauri Halderman, in particular, gently but sternly reminded me when I left my position at AMNH to never lose my writer's voice. I also received encouragement in science writing from the mentorship of Richard Panek.

This research would not have been possible without funding. I am grateful to the Michael Kraus Research Grant from the American Historical Association, the Baird Society Resident Fellowship from the Smithsonian Institution Libraries, the Jay T. Last Fund Research Fellowship from the American Antiquarian Society, the American Society for Eighteenth-Century Studies Fellowship from the Boston Athenaeum, the Fothergill Research Award from the Bartram Trail Conference, the Deakin-Royce Graduate Research Fellowship in Australian Studies from Harvard University, the Jacob M. Price Visiting Research Fellowship from the William L. Clements Library at the University of Michigan, a travel grant from the Lewis Walpole Library at Yale University, a travel fellowship from the Linda Hall Library, a summer research grant from the Charles Warren Center for Studies in American History at Harvard, the Jens Aubrey Westengard Fellowship from Harvard, a research fellowship from the Consortium for History of Science, Technology, and Medicine, fellowships from Harvard's Graduate School of Arts and Sciences, and several awards that let me attend conferences to receive feedback on the project from the American Studies Program at Harvard, the Center for American Political Studies at Harvard, and the National Science Foundation. I especially thank Dartmouth College for research funds to offset image permissions and a subvention to cover book production costs.

Several of these grants came from libraries, archives, and museums; many other librarians, archivists, and curators have been immensely generous with their time and expertise. I am grateful to staff at the American Antiquarian Society (especially Nan Wolverton), the Boston Athenaeum (especially Carolle Morini), the American Philosophical Society, the Historical Society of Pennsylvania, the Academy of Natural Sciences of Drexel University (especially Jennifer Vess and Mark Sabaj), the Wellcome Collection, the Royal Society (especially Virginia Mills), the Natural History Museum in London (especially Patrick Campbell, Miranda Lowe, and James Maclaine), the Harvard University Archives, the Collection of Historical Scientific Instruments (especially Sara Schechner), Houghton Library, the Francis A. Countway Library of Medicine, the Harvard University Herbaria, the Clendening History of Medicine Library at the University of Kansas Medical Center, the Santa Barbara Museum of Natural History, the Schlesinger Library on the History of Women in America, Harvard's Museum of Comparative Zoology (especially Karsten E. Hartel, Joe Martinez, and Mark Omura), the Linnean Society of London (especially Lynda Brooks, Isabelle Charmantier, and Ollie Crimmen), the William L. Clements Library (especially Jane Ptolemy), the Museum De Lakenhal (especially Bart Doff and Rob Wolthoorn), Special Collections at the University of California Libraries (especially Melinda Hayes and Marje Schuetze-Coburn), the Dibner Library of the History of Science and Technology at the Smithsonian (especially Morgan Aronson and Lilla Vekerdy), the Museum für Naturkunde Berlin (especially Edda Aßel), the Joseph F. Cullman Library at the Smithsonian (especially Leslie Overstreet and Allie Newman), the Lewis Walpole Library (especially Nicole Bouché, Scott Poglitsch, and Susan Walker), the Linda Hall Library (especially the staff's willingness to permit special collections research on a snow day), the Library of Congress, the Gilcrease Museum, the Bodleian Library at the University of Oxford, the State Library of New South Wales (especially Wendy Holz), and the National Library of Australia. Fred Burchsted at Harvard University deserves special thanks for his wealth of information about research resources and early American natural history. I am also grateful for honors from the Hakluyt Society Essay Prize and Harvard University's Bowdoin Prize in the Natural Sciences.

This book likewise owes a debt to dozens of random acts of kindness. The librarians listed above helped me locate sources, sort through image permissions, and look at fish scales. More than once, I've been guided by a man on a ship. And numerous herpetologists, ichthyologists, mammologists, and other scientists showed me, a complete stranger, their specimen holdings or answered questions in person, over email, or by phone. They followed up with additional journal articles and leads that I never would have found without their help. John Corbett graciously took me to the Great Barrier Reef (I would later learn it was his birthday!) for several hours, while Shane Marks guided us through the reef's ecology. Debbie Corbett offered hospitality during our stay in Far North Queensland. After my trip, Aaron Hartmann took time to sit down with me and examine photographs I took while snorkeling on Endeavour Reef. He gave advice for assessing the health of coral reefs and vetted the polyp diagram expertly designed by Emily Damstra, to whom I am also indebted. Rulon Clark, Aaron Place, and Corey Fincher all provided resources and answered questions as I explored the complex behaviors of rattlesnakes. Brendan Clifford took a full day out of his busy life to go search for rattlesnakes with me during the double whammy of a pandemic and dangerous heat wave. Joe Higgins taught me how to make fresh fish prints. Carlos drove me to an isolated Mexican port to find raccoons. And Suzanne MacDonald graciously answered questions about raccoon intelligence.

Any errors that remain are my own.

Finally, I would be remiss in a book of this breed if I did not thank the many animals who helped the project along the way. From the dwarf raccoons who made good company in Cozumel, to the surprising persistence of a long-lost rattlesnake, to the coral polyps beneath my fins, to a fish from the grocery store who gave its life to the history of science: this has been their story, and for them, I am grateful.

NOTES

INTRODUCTION

1. James Petiver, *Brief Directions for the Easie Making, and Preserving Collections of all Natural Curiosities* (London: c. 1715). For a shark found with "a human corpse in his belly," for instance, see Oliver Goldsmith, *An History of the Earth, and Animated Nature* (London: J. Nourse, 1774), vol. 6: 238. On Petiver's reliance on the slave trade, see Kathleen S. Murphy, "Collecting Slave Traders: James Petiver, Natural History, and the British Slave Trade," *William and Mary Quarterly* 70, no. 4 (October 2013): 637–670. See also Marcus Rediker, "History from Below the Water Line: Sharks and the Atlantic Slave Trade," *Atlantic Studies* 5, no. 2 (August 2008): 285–297. For one powerful meditation on the role of predators in human history and thought, see David Quammen, *Monster of God: The Man-Eating Predator in the Jungles of History and the Mind* (New York: W. W. Norton, 2003).
2. Mark Sabaj, email conversation with the author, March 6, 2022. For species discovered in stomachs and the meanings behind their names, see Christopher Scharpf, "Name of the Week 2021: Holotypes 'Collected' by Other Animals," ETYFish Project, September 1, 2021, https://etyfish.org/name-of-the-week2021/ (accessed March 15, 2022).
3. On the impact of animals—particularly mammals—on colonialism, see Virginia DeJohn Anderson, *Creatures of Empire: How Domestic Animals Transformed Early America* (Oxford: Oxford University Press, 2004); Andrea L. Smalley, *Wild by Nature: North American Animals Confront Colonization* (Baltimore: Johns Hopkins University Press, 2017). On the challenges of animal and specimen transport, see also Christopher M. Parsons and Kathleen S. Murphy, "Ecosystems Under Sail: Specimen Transport in Eighteenth-Century French and British Atlantics," *Early American Studies* 10, no. 3 (Fall 2012): 503–539; Louise E. Robbins, *Elephant Slaves*

and Pampered Parrots: Exotic Animals in Eighteenth-Century Paris (Baltimore: Johns Hopkins University Press, 2002).

4. Daines Barrington, *Miscellanies* (London: J. Nichols, 1781), 163.
5. "Curiosity," in N. [Nathaniel] Bailey, *An Universal Etymological English Dictionary* . . . (London: 1763 [1721]), unpaginated. See also Susan Scott Parrish, *American Curiosity: Cultures of Natural History in the Colonial British Atlantic World* (Chapel Hill: Omohundro Institute of Early American History and Culture and University of North Carolina Press, 2006); Philip Ball, *Curiosity: How Science Became Interested in Everything* (Chicago: University of Chicago Press, 2012); Barbara M. Benedict, *Curiosity: A Cultural History of Early Modern Inquiry* (Chicago: University of Chicago Press, 2001).
6. Georges-Louis Leclerc, Comte de Buffon, *Natural History, General and Particular* . . . , trans. William Smellie, 3rd ed. (London: A. Strahan and T. Cadell, 1791), vol. 3: 303.
7. For one instance of this standard picture, see James Delbourgo, *Collecting the World: The Life and Curiosity of Hans Sloane* (London: Allen Lane, 2017), esp. xxvi–xxvii and 121. For works suggesting the limits of human understanding in Enlightenment scientific study, see Jennifer L. Anderson, *Mahogany: The Costs of Luxury in Early America* (Cambridge: Harvard University Press, 2012), esp. 249; Caroline Winterer, *American Enlightenments: Pursuing Happiness in the Age of Reason* (New Haven: Yale University Press, 2016), esp. 17; Jan Golinski, *British Weather and the Climate of Enlightenment* (Chicago: University of Chicago Press, 2007), esp. 10; Michael Gaudio, "Swallowing the Evidence: William Bartram and the Limits of Enlightenment," *Winterthur Portfolio* 36, no. 1 (2001): 1–17. On cultural opposition to Enlightenment philosophes, see Darrin M. McMahon, *Enemies of the Enlightenment: The French Counter-Enlightenment and the Making of Modernity* (Oxford: Oxford University Press, 2002). For one critique of conceiving of "the Enlightenment" as a singular and unified front, see J. G. A. Pocock, "Historiography and Enlightenment: A View of Their History," *Modern Intellectual History* 5, no. 1 (2008): 83–96.
8. See also David Gary Shaw, "A Way with Animals," *History and Theory* 52, no. 4 (December 2013): 1–12. For an early monograph centering animals as historical actors, see Harriet Ritvo, *The Animal Estate: The English and Other Creatures in the Victorian Age* (Cambridge: Harvard University Press, 1987). For my fuller thoughts on the subject of nonhuman agency, see Whitney Barlow Robles, "On Nonhuman Agency," *Journal of Interdisciplinary History* (forthcoming). See also Whitney Barlow Robles, "The Rattlesnake and the Hibernaculum: Animals, Ignorance, and Extinction in the Early American Underworld," *William and Mary Quarterly* 78, no. 1 (January 2021): 3–44, esp. 9–12. In brief, I agree with scholars who decouple agency and intention (as with Timothy Mitchell and Jane Bennett); I agree with scholars who situate the term "agency" historically (as with Jessica Riskin and Erica Fudge); I agree with scholars who turn our sights toward the vast multiplicity of nonhuman forces at hand in human history (as with Paul Sutter); and I especially agree with scholars who fault the latest scholarly discourse for eliding long-

standing Indigenous theorizations of nonhuman agency (as with Zoe Todd, Kim TallBear, and Bathsheba Demuth). See Timothy Mitchell, "Can the Mosquito Speak?" in *Rule of Experts: Egypt, Techno-Politics, Modernity* (Berkeley: University of California Press, 2002); Jane Bennett, *Vibrant Matter: A Political Ecology of Things* (Durham, N.C.: Duke University Press, 2010); Jessica Riskin, *The Restless Clock: A History of the Centuries-Long Argument over What Makes Living Things Tick* (Chicago: University of Chicago Press, 2016), esp. 3; Erica Fudge, "What Was It Like to Be a Cow?: History and Animal Studies," in *The Oxford Handbook of Animal Studies*, ed. Linda Kalof (Oxford: Oxford University Press, 2014); Paul S. Sutter, "The World with Us: The State of American Environmental History," *Journal of American History* 100, no. 1 (June 2013): esp. 98; Zoe Todd, "An Indigenous Feminist's Take on the Ontological Turn: 'Ontology' Is Just Another Word for Colonialism," *Journal of Historical Sociology* 29 (March 2016): 4–22; Kim TallBear, "Beyond the Life/Not-Life Binary: A Feminist-Indigenous Reading of Cryopreservation, Interspecies Thinking, and New Materialisms," in *Cryopolitics: Frozen Life in a Melting World*, ed. Joanna Radin and Emma Kowal (Cambridge: MIT Press, 2017); Bathsheba Demuth, *Floating Coast: An Environmental History of the Bering Strait* (New York: W. W. Norton, 2019).

9. For one influential critique of presentism, see Lynn Hunt, "From the President: Against Presentism," *Perspectives on History* (May 2002), www.historians.org/publications-and-directories/perspectives-on-history/may-2002/against-presentism (accessed March 15, 2022). See also Daniel Steinmetz-Jenkins, "Beyond the End of History: Historians' Prohibition on 'Presentism' Crumbles Under the Weight of Events," *Chronicle of Higher Education* (August 14, 2020), www.chronicle.com/article/beyond-the-end-of-history?cid=gen_sign_in (accessed March 15, 2022).

10. On writing like an oyster, see Tamara Fernando, "Seeing Like the Sea: A Multispecies History of the Ceylon Pearl Fishery, 1800–1925," *Past and Present* 254, no. 1 (February 2022): 127–160, esp. 133. Anthropologists, of course, have long wrestled with questions of embodied experience versus analytic distance, especially in studies of nonhumans and the environment. See in particular Anna Lowenhaupt Tsing, *The Mushroom at the End of the World: On the Possibility of Life in Capitalist Ruins* (Princeton: Princeton University Press, 2015); Eduardo Kohn, *How Forests Think: Toward an Anthropology Beyond the Human* (Berkeley: University of California Press, 2013); Hugh Raffles, *Insectopedia* (New York: Vintage, 2010). On using scientific studies in historical research, see also Whitney Barlow Robles, "Science for the History of Science: An Imperfect Tool," *Uncommon Sense* (Omohundro Institute of Early American History and Culture blog), March 3, 2021, https://blog.oieahc.wm.edu/science-for-the-history-of-science-an-imperfect-tool/ (accessed March 15, 2022).

11. Meghan K. Roberts, *Sentimental Savants: Philosophical Families in Enlightenment France* (Chicago: University of Chicago Press, 2016), 8. See also Jessica Riskin, *Science in the Age of Sensibility: The Sentimental Empiricists of the French Enlightenment* (Chicago: University of Chicago Press, 2002).

12. Nathan Pacoureau et al., "Half a Century of Global Decline in Oceanic Sharks and Rays," *Nature* 589 (2021): 567–571.
13. "Animal," *Merriam-Webster.com Dictionary,* Merriam-Webster, www.merriam-webster.com/dictionary/animal (accessed August 3, 2022).
14. On the value of species-specific and microhistorical approaches to the history of science, see also Marcy Norton, "The Quetzal Takes Flight: Microhistory, Mesoamerican Knowledge, and Early Modern Natural History," in *Translating Nature: Cross-Cultural Histories of Early Modern Science,* ed. Jaime Marroquín Arredondo and Ralph Bauer (Philadelphia: University of Pennsylvania Press, 2019), 119–147; Iris Montero Sobrevilla, "The Slow Science of Swift Nature: Hummingbirds and Humans in New Spain," in *Global Scientific Practice in an Age of Revolutions, 1750–1850* (Pittsburgh: University of Pittsburgh Press, 2016), 127–146.
15. Letter from David Skene to John Ellis, May 16, 1765, John Ellis Manuscripts, vol. 2, Linnean Society of London.
16. For the most influential intellectual history of the concept, see Arthur O. Lovejoy, *The Great Chain of Being: A Study of the History of an Idea* (New York: Harper and Row, 1960 [1936]). Lovejoy writes: "It was in the eighteenth century that the conception of the universe as a Chain of Being, and the principles which underlay this conception—plenitude, continuity, gradation—attained their widest diffusion and acceptance. . . . There has been no period in which writers of all sorts—men of science and philosophers, poets and popular essayists, deists and orthodox divines—talked so much about the Chain of Being" (183).
17. Soame Jenyns, *The Works of Soame Jenyns . . .* (London: T. Cadell, 1790), vol. 3: 179–180, 184; Lovejoy, *The Great Chain of Being,* 198–199.
18. Jenyns, *The Works of Soame Jenyns,* vol. 3: 184–185. On humanity as a still-unsettled category in the early modern period, see also Sylvia Wynter, "Unsettling the Coloniality of Being/Power/Truth/Freedom: Towards the Human, After Man, Its Overrepresentation—An Argument," *CR: The New Centennial Review* 3, no. 3 (Fall 2003): 257–337; Erica Fudge, *Brutal Reasoning: Animals, Rationality, and Humanity in Early Modern England* (Ithaca: Cornell University Press, 2006); Lynn Festa, *Fiction Without Humanity: Person, Animal, Thing in Early Enlightenment Literature and Culture* (Philadelphia: University of Pennsylvania Press, 2019).
19. For a brief sampling of scholarship foregrounding female, enslaved, Native, and lay actors in early modern natural history, see Parrish, *American Curiosity;* Londa Schiebinger, *Plants and Empire: Colonial Bioprospecting in the Atlantic World* (Cambridge: Harvard University Press, 2004); Andrew J. Lewis, *A Democracy of Facts: Natural History in the Early Republic* (Philadelphia: University of Pennsylvania Press, 2011); Cameron B. Strang, *Frontiers of Science: Imperialism and Natural Knowledge in the Gulf South Borderlands, 1500–1850* (Chapel Hill: Omohundro Institute of Early American History and Culture and University of North Carolina Press, 2018); Kathleen S. Murphy, "Translating the Vernacular: Indigenous and African Knowledge in the Eighteenth-Century British Atlantic," *Atlantic Studies* 8, no. 1 (2011): 29–48. For a contemporary investigation of interlocking systems of oppression that reach across the human-animal divide, see Aph Ko and Syl Ko,

Aphro-Ism: Essays on Pop Culture, Feminism, and Black Veganism from Two Sisters (Brooklyn, N.Y.: Lantern, 2017), esp. 27.

20. On the history of fluid preservation of specimens, see John E. Simmons, *Fluid Preservation: A Comprehensive Reference* (Lanham, Md.: Rowman & Littlefield, 2014).
21. Sven Kullander, "Museum Adolphi Friderici," http://linnaeus.nrm.se/zool/madfrid.html.en (accessed August 9, 2022); Wilfrid Blunt, *Linnaeus: The Compleat Naturalist* (Princeton: Princeton University Press, 2001), 211.
22. Samuel J. Redman, "Bodies of Knowledge: Philadelphia and the Dark History of Collecting Human Remains," *Perspectives on History*, September 15, 2022, www.historians.org/research-and-publications/perspectives-on-history/october-2022/bodies-of-knowledge-philadelphia-and-the-dark-history-of-collecting-human-remains (accessed October 18, 2022); Justin Dunnavant, Delande Justinvil, and Chip Colwell, "Craft an African American Graves Protection and Repatriation Act," *Nature* 593 (May 20, 2021): 337–340; Chip Colwell, *Plundered Skulls and Stolen Spirits: Inside the Fight to Reclaim Native America's Culture* (Chicago: University of Chicago Press, 2017). For a sampling of recent battles over Black heritage and human remains, see Krystal Strong, "A Requiem for Delisha and Tree Africa," *Anthropology News* 62, no. 5 (September/October 2021), www.anthropology-news.org/articles/a-requiem-for-delisha-and-tree-africa/ (accessed October 1, 2022); Valentina Di Liscia, "Legal Precedents or Reparations? Lawsuit Against Harvard May Decide Who Owns Images of Enslaved People," *Hyperallergic*, October 27, 2021, www.hyperallergic.com/687964/lawsuit-against-harvard-may-decide-who-owns-images-of-enslaved-people/ (accessed October 1, 2022).
23. On Linnaean naming, see also Stephen B. Heard, *Charles Darwin's Barnacle and David Bowie's Spider: How Scientific Names Celebrate Adventurers, Heroes, and Even a Few Scoundrels*, illustrated by Emily S. Damstra (New Haven: Yale University Press, 2020).
24. For Kwasi's story, see Parrish, *American Curiosity*, 1–6. On the formation of the discipline of natural history in the eighteenth century and its continued impact on the modern life sciences, see also Paul Lawrence Farber, *Finding Order in Nature: The Naturalist Tradition from Linnaeus to E. O. Wilson* (Baltimore: Johns Hopkins University Press, 2000).
25. Juan Pimentel, *The Rhinoceros and the Megatherium: An Essay in Natural History*, trans. Peter Mason (Cambridge: Harvard University Press, 2017). On natural history and nature studies during the Renaissance, see also Brian W. Ogilvie, *The Science of Describing: Natural History in Renaissance Europe* (Chicago: University of Chicago Press, 2006); Paula Findlen, *Possessing Nature: Museums, Collecting, and Scientific Culture in Early Modern Italy* (Berkeley: University of California Press, 1994); Mackenzie Cooley, *The Perfection of Nature: Animals, Breeding, and Race in the Renaissance* (Chicago: University of Chicago Press, 2022).
26. Keith Thomas, *Man and the Natural World: Changing Attitudes in England, 1500–1800* (Oxford: Oxford University Press, 1996 [1983]), esp. 52; Michel Foucault, *The Order of Things: An Archaeology of the Human Sciences* (New York: Vintage, 1994

[1966]), esp. 131. On the rise of objectivity in the nineteenth century, see Lorraine Daston and Peter Galison, *Objectivity* (Brooklyn, N.Y.: Zone Books, 2007).

27. See, for instance, Carl Safina, *Eye of the Albatross: Visions of Hope and Survival* (New York: Henry Holt, 2002); Sy Montgomery, *The Soul of an Octopus: A Surprising Exploration into the Wonder of Consciousness* (New York: Atria Paperback, 2015).
28. Verlyn Klinkenborg, *Timothy; or, Notes of an Abject Reptile* (New York: Alfred A. Knopf, 2006).

CHAPTER 1. A DIFFERENT SPECIES OF RESISTANCE

1. James Cook, June 22, 1770, *The Journals of Captain James Cook on His Voyages of Discovery: The Voyage of the* Endeavour *1768–1771*, ed. J. C. Beaglehole (Cambridge: Cambridge University Press for the Hakluyt Society, 1955), vol. 1: 350; Sydney Parkinson, *A Journal of a Voyage to the South Seas . . .* (London: Stanfield Parkinson, 1773), 143.
2. James Edward Smith, *An Introduction to Physiological and Systematical Botany* (London: Longman, Hurst, Rees, and Orme, 1807), 4.
3. Mark J. A. Vermeij et al., "Coral Larvae Move Toward Reef Sounds," *PLoS ONE* 5, no. 5 (May 14, 2010): e10660; Sylvie Tambutté et al., "Coral Biomineralization: From the Gene to the Environment," *Journal of Experimental Marine Biology and Ecology* 408, no. 1–2 (November 2011): 59.
4. Thomas Hobbes, *Leviathan; or, The Matter, Forme, and Power of a Common-Wealth, Ecclesiasticall and Civil* (London: Andrew Crooke, 1651), 40.
5. John Ellis, *An Essay Towards a Natural History of the Corallines, and Other Marine Productions of the Like Kind . . .* (London: 1755), 78.
6. For scholarship on the ambiguous classification and visual depiction of corals and other zoophytes, see Susannah Gibson, *Animal, Vegetable, Mineral? How Eighteenth-Century Science Disrupted the Natural Order* (Oxford: Oxford University Press, 2015); Barbara M. Stafford, "Images of Ambiguity: Eighteenth-Century Microscopy and the Neither/Nor," in *Visions of Empire: Voyages, Botany, and Representations of Nature*, ed. David Philip Miller and Peter Hanns Reill (Cambridge: Cambridge University Press, 1996), 230–257; Elizabeth Athens, "Chaotic Life: Representing the Freshwater Polyp," *Journal18* (August 2016), www.journal18.org/774 (accessed September 7, 2022). On Darwin's "coral of life" comparison, see Charles Darwin, "Notebook B [1837–1838]," *Charles Darwin's Notebooks, 1836–1844: Geology, Transmutation of Species, Metaphysical Enquiries*, ed. Paul H. Barrett, Peter J. Gautrey, Sandra Herbert, David Kohn, and Sydney Smith (Ithaca: Cornell University Press, 1987), 177; Horst Bredekamp, *Darwins Korallen: Frühe Evolutionsmodelle und die Tradition der Naturgeschichte* (Berlin: Verlag Klaus Wagenbach, 2005).
7. [Paracelsus], *Paracelsus His Dispensatory and Chirurgery . . .* , trans. W. D. (London: Printed by T. M. for Philip Chetwind, 1656), 41. On coral and Ovid, see especially Michael Cole, "Cellini's Blood," *Art Bulletin* 81, no. 2 (June 1999): 228–230; J. Malcolm Shick, *Where Corals Lie: A Natural and Cultural History* (London: Reaktion Books, 2018), 77–80. On the material culture of Renaissance red coral, see Marion

Endt-Jones, "Coral Shrine from Trapani (c. 1650)," in *Coral: Something Rich and Strange*, ed. Marion Endt-Jones (Liverpool: Liverpool University Press, 2013), 59.

8. Joseph Banks, *The Endeavour Journal of Joseph Banks, 1768–1771*, ed. John C. Beaglehole (Sydney, 1962), vol. 1: 363–364; Alfio Ferrara, "On the Coral Fishery in the Sicilian Seas," *A Journal of Natural Philosophy, Chemistry, and the Arts* 33 (October 1812): 136–145; Gilbert Buti, "Du Rouge pour le Noir: du Corail Méditerranéen pour la Traite Négrière au xviiie Siècle," *Rives Méditerranéennes* 57, no. 2 (2018): 109–127.
9. Jacques-François Dicquemare, "On Sea Netles," November 1772, Archive Papers (1768–1780), Royal Society Archives, London; Goldsmith, *An History of the Earth*, vol. 8: 165; Dicquemare, "A Third Essay on Sea-Anemonies," *Philosophical Transactions* 67 (1777): 58.
10. J. F. Gronovius, "Extract of a Letter from J. F. Gronovius M.D. at Leyden to Peter Collinson F.R.S. Concerning a *Water-Insect* wch Being Cut into Severall Peices Becomes *So Many Perfect Animals*," Letters and Papers Volume 2 Decade 1 (1742), Royal Society Archives. Gronovius's letter was read at the Royal Society in November of 1742 and later printed, in amended form, in the *Philosophical Transactions*. Abraham Trembley, "Observations and Experiments upon the Freshwater Polypus, by Monsieur Trembley, at the Hague, Translated from the French by P.H.Z. F.R.S.," *Philosophical Transactions* 42 (1742–1743): v and x; William Frederic Martyn, *A New Dictionary of Natural History; or, Compleat Universal Display of Animated Nature, with Accurate Representations of the Most Curious and Beautiful Animals, Elegantly Coloured* (London: Harrison, 1785), vol. 2, unpaginated.
11. Lazzaro Spallanzani, *Tracts on the Natural History of Animals and Vegetables . . .*, trans. John Graham Dalyell, 2nd ed. (Edinburgh: William Creech and Archd. Constable, 1803), vol. 1: 178. For one classic argument showing how human intentions never unfold in a vacuum but are embedded in and partly produced by their environments, see Linda Nash, "The Agency of Nature or the Nature of Agency?" *Environmental History* 10, no. 1 (January 2005): 67–69.
12. Letter from David Skene to John Ellis, December 16, 1768, John Ellis Manuscripts, vol. 2, Linnean Society of London; William Stukeley, "On Corals, Corallines &c: a Letter to Mr Collison [*sic*] May 1752," 1752, Wellcome Library, MS 4728 ("curious windings").
13. See Buffon, *Natural History, General and Particular . . .*, trans. William Smellie, 2nd ed. (London: W. Strahan and T. Cadell, 1785), vol. 2: 8. For one account of the Venus flytrap and Ellis's study of it, see E. Charles Nelson, *Aphrodite's Mousetrap: A Biography of Venus's Flytrap* (Aberystwyth, Wales: Boethius Press, 1990). On Ellis's use of burning and chemical tests to distinguish animal and plant natures, see Susannah Gibson, "On Being an Animal, Or, the Eighteenth-century Zoophyte Controversy in Britain," *History of Science* 50, no. 4 (2012): 453–476.
14. James Parsons, *Philosophical Observations on the Analogy Between the Propagation of Animals and that of Vegetables . . .* (London: C. Davis, 1752), 201, 219–222. For scholarly accounts that investigate early modern critiques of Cartesianism as

applied to animals and show misunderstandings of the actual arguments of Descartes, see Virginia P. Dawson, "The Problem of Soul in the 'Little Machines' of Réaumur and Charles Bonnet," *Eighteenth-Century Studies* 18, no. 4 (Autumn 1985): 503–522; Heather Keenleyside, *Animals and Other People: Literary Forms and Living Beings in the Long Eighteenth Century* (Philadelphia: University of Pennsylvania Press, 2016); Anita Guerrini, *The Courtiers' Anatomists: Animals and Humans in Louis XIV's Paris* (Chicago: University of Chicago Press, 2015), 3–4; Riskin, *The Restless Clock,* 45, 68, and 75.

15. Goldsmith, *An History of the Earth,* vol. 2: 2, and vol. 8: 163; William Smellie, *The Philosophy of Natural History* (Dublin: William Porter, 1790), vol. 1: 5.
16. Henry Baker, *An Attempt Towards a Natural History of the Polype . . .* (London: R. Dodsley, 1743), 100, 81, 68, 48.
17. Baker, *Natural History of the Polype,* 9.
18. Baker, *Natural History of the Polype,* 204. On philosophical controversies surrounding Trembley's polyp, see Virginia P. Dawson, *Nature's Enigma: The Problem of the Polyp in the Letters of Bonnet, Trembley, and Réaumur* (Philadelphia: American Philosophical Society, 1987).
19. Baker, *Natural History of the Polype,* 204–205, 209.
20. On the move from freshwater to saltwater in studies of regeneration, see also Mary Terrall, *Catching Nature in the Act: Réaumur and the Practice of Natural History in the Eighteenth Century* (Chicago: University of Chicago Press, 2014), 124–125.
21. William Borlase to Ellis, 1755, John Ellis Manuscripts, vol. 1, Linnean Society of London. Borlase would not remain so optimistic. Several years later, in the thick of the Seven Years' War, he demurred to Ellis: "I do not at all wonder that Nat. History is neglected—our all is in danger, our leaders by the ears, war abroad and almost famine at home." See Borlase to Ellis, undated, c. 1759, John Ellis Manuscripts, vol. 1, Linnean Society of London. On Peyssonnel, see Jan Vandersmissen, "Experments and Evolving Frameworks of Scientific Exploration: Jean-André Peyssonnel's Work on Coral," in *Expeditions as Experiments: Practising Observation and Documentation,* ed. Marianne Klemun and Ulrike Spring (London: Palgrave Macmillan, 2016).
22. On the deep-sea octocoral, see Ellis, *Essay Towards a Natural History of the Corallines,* 96; Gary C. Williams, "The Global Diversity of Sea Pens (Cnidaria: Octocorallia: Pennatulacea)," *PLoS One* (2011): e22747.
23. George Humphrey, manuscript handbook, "Directions for Collecting and Preserving All Kinds of Natural Curiosities," 1776, Collection 371, Academy of Natural Sciences, 61; John Ellis to Carl Linnaeus, August 19, 1768, in *A Selection of the Correspondence of Linnaeus, and Other Naturalists: From the Original Manuscripts,* ed. James Edward Smith (London: Longman, Hurst, Rees, Orme, and Brown, 1821), vol. 1: 231. On the damage caused to the Great Barrier Reef, in particular, from centuries of coral collecting, see Ben Daley and Peter Griggs, "'Loved to Death': Coral Collecting in the Great Barrier Reef, Australia, 1770–1970," *Environment and History* 14, no. 1 (February 2008): 89–119.

24. Humphrey, "Directions for Collecting and Preserving All Kinds of Natural Curiosities," 109; Dicquemare, "On Sea Netles."
25. William Brownrigg to Ellis, John Ellis Manuscripts, vol. 1, Linnean Society of London; Jean-André Peyssonnel, "An Account of a Manuscript Treatise Presented to the Royal Society . . . Upon Coral, and Several Other Productions Furnish'd by the Sea, in Order to Illustrate the Natural History Thereof . . . Extracted and Translated from the French by Mr. William Watson, F.R.S.," *Philosophical Transactions* 47 (1752): 447–448.
26. Mordach [Murdoch] Mackenzie to Ellis, November 15, 1756, and January 3, 1758, John Ellis Manuscripts, vol. 2, Linnean Society of London; Banks, *The Endeavour Journal of Joseph Banks,* vol. 2: 108.
27. Kevin Dawson, *Undercurrents of Power: Aquatic Culture in the African Diaspora* (Philadelphia: University of Pennsylvania Press, 2018), 2.
28. Letter from Alexander Garden to John Ellis, March 25, 1755, in *A Selection of the Correspondence of Linnaeus,* vol. 1: 349.
29. George Humphrey, "Directions for Collecting and Preserving All Kinds of Natural Curiosities," 107. On enslaved and Indigenous divers and fishers as collectors of corals, pearls, sunken treasure, and other underwater curiosities throughout the Americas, see Parrish, *American Curiosity,* 271–274; Dawson, "History from Below: Enslaved Underwater Divers," in *Undercurrents of Power,* 57–84; James Delbourgo, "Divers Things: Collecting the World Under Water," *History of Science* 49, no. 163 (June 2011): 149–185; Molly A. Warsh, "A Political Ecology in the Early Spanish Caribbean," *William and Mary Quarterly* 71, no. 4 (October 2014): 517–548; Rebekka von Mallinckrodt, "Exploring Underwater Worlds: Diving in the Late Seventeenth-/Early Eighteenth-Century British Empire," in *Empire of the Senses: Sensory Practices of Colonialism in Early America,* ed. Daniela Hacke and Paul Musselwhite (Leiden and Boston: Brill, 2017), 300–322.
30. Griffith Hughes, *The Natural History of Barbados* (London: 1750), 294–296.
31. Hughes, *The Natural History of Barbados,* 293; Griffith Hughes, "A Letter from the Rev[d] Mr. Griffith Hughes, Minister of St. Lucy's Parish in Barbadoes, to Martin Folkes, Esq; Pr. R.S. Concerning a Zoophyton, Somewhat Resembling the Flower of the Marigold," *Philosophical Transactions* 42 (1743): 590.
32. Carl Linnaeus to John Ellis, November 8, 1769, in *A Selection of the Correspondence of Linnaeus,* vol. 1: 240.
33. "Colony, n.," OED Online, June 2022, Oxford University Press, www-oed-com.dartmouth.idm.oclc.org/view/Entry/36547?rskey=kcaYFf&result=1&isAdvanced=false (accessed August 31, 2022); "colony, n.s.," in Samuel Johnson, *A Dictionary of the English Language,* 1755, Johnson's Dictionary Online, https://johnsonsdictionaryonline.com/1755/colony_ns (accessed August 31, 2022); John Hunter, "Observations on Bees," *Philosophical Transactions* 82 (1792): 141; Goldsmith, *An History of the Earth,* vol. 2: 25; Erasmus Darwin, "Additional Notes," in *The Botanic Garden: A Poem, In Two Parts,* 3rd ed. (London: J. Johnson, 1795 [1791]), 49.
34. Peyssonnel, "An Account of a Manuscript Treatise Presented to the Royal Society," 467; John Ellis, "A Letter from Mr. John Ellis, F.R.S. to Mr. Peter Collinson, F.R.S.

Concerning the Animal Life of Those Coral-Lines, That Look Like Minute Trees, and Grow upon Oysters and Fucus's All Round the Sea-Coast of This Kingdom," *Philosophical Transactions* 48 (1754): 631.

35. John Albert Schlosser, "An Account of a Curious, Fleshy, Coral-Like Substance; in a Letter to Mr. Peter Collinson, F.R.S. from Dr. John Albert Schlosser, M.D. F.R.S. with Some Observations on It Communicated to Mr. Collinson by Mr. John Ellis, F.R.S.," *Philosophical Transactions* 49 (1756): 451; William Whewell, *Of the Plurality of Worlds: A Facsimile of the First Edition of 1853 . . .* , ed. Michael Ruse (Chicago: University of Chicago Press, 2001), 24. Although Hobbes, in the seventeenth century, did not compare his leviathan to coral, animal bodies and naturalistic analogies and metaphors were crucial to his thinking. Indeed, the introduction to *Leviathan* opens with the word "nature": "NATURE (the Art whereby God hath made and governes the World) is by the *Art* of man, as in many other things, so in this also imitated, that it can make an Artificial Animal. . . . That great LEVIATHAN called a COMMON-WEALTH, or STATE . . . is but an Artificiall Man." See Hobbes, *Leviathan,* 1.
36. Letter from Alexander Garden to John Ellis, March 25, 1755, in *A Selection of the Correspondence of Linnaeus,* vol. 1: 349; Ellis, *Essay Towards a Natural History of the Corallines,* 78 and 103.
37. Ellis, *Essay Towards a Natural History of the Corallines,* 102.
38. John Ellis, "An Account of the Actinia Sociata, or Clustered Animal-Flower, Lately Found on the Sea-Coasts of the New-Ceded Islands," *Philosophical Transactions* 57 (1767): 433. *Actinia* is a genus of sea anemones, which were often called "animal flowers" at the time, but Ellis's *Actinia sociata* is now named *Zoanthus sociatus* and is a polyp-based zoanthid (also part of the phylum Cnidaria).
39. Benjamin Franklin, *Observations Concerning the Increase of Mankind . . .* (Boston: S. Kneeland, 1755), 3, 13–15. See also Joyce E. Chaplin, *Benjamin Franklin's Political Arithmetic: A Materialist View of Humanity* (Washington, D.C.: Smithsonian Institution Libraries, 2006).
40. Bathsheba Demuth, "The Walrus and the Bureaucrat: Energy, Ecology, and Making the State in the Russian and American Arctic, 1870–1950," *American Historical Review* 124, no. 2 (April 2019): 487.
41. Parkinson, *A Journal of a Voyage to the South Seas,* 142; Banks, *The Endeavour Journal of Joseph Banks,* vol. 2: 81.
42. For Stukeley's telling of Newton's anecdote, see William Stukeley, "Memoirs of Sr. Isaac Newtons Life," MS/142, Royal Society Archives, 15–16.
43. Stukeley, "On Corals, Corallines &c."
44. Lovejoy, *The Great Chain of Being,* 183; Stukeley, "On Corals, Corallines &c." On the business and commerce of natural history, including botanical gardens, see Dániel Margócsy, *Commercial Visions: Science, Trade, and Visual Culture in the Dutch Golden Age* (Chicago: University of Chicago Press, 2014).
45. Stukeley, "On Corals, Corallines &c."; James Parsons, "A Letter from James Parsons, M.D. F.R.S. to the Rev. Mr. Birch, Secr. R.S. Concerning the Formation of Corals, Corallines, &c.," *Philosophical Transactions* 47 (1752): 511–512.

46. Stukeley, "On Corals, Corallines &c."
47. William Stukeley, May 7, 1752, *The Family Memoirs of the Rev. William Stukeley, M.D. . . .* (London: Surtees Society, 1883), vol. 2: 377; Stukeley, "On Corals, Corallines &c."
48. Parsons, "A Letter from James Parsons, M.D. F.R.S. to the Rev. Mr. Birch," 505; Stukeley, "On Corals, Corallines &c."
49. Mordach [Murdoch] Mackenzie to Ellis, June 2, 1756, John Ellis Manuscripts, vol. 2, Linnean Society of London; William Bartram, "William Bartram Notes on the Nature of Animals with Sketches, 1780s," Bartram Family Papers, Historical Society of Pennsylvania, Collection 36, Box 1, Folder 83.
50. John Reinhold Forster, *Observations Made During a Voyage Round the World . . .* (London: G. Robinson, 1778), 148–151. For more on the reef theories of Forster, Flinders, and Darwin, see Alistair Sponsel, "From Cook to Cousteau: The Many Lives of Coral Reefs," in *Fluid Frontiers: Exploring Oceans, Islands, and Coastal Environments,* ed. John Gillis and Franziska Torma (Cambridge, U.K.: White Horse Press, 2015).
51. Goldsmith, *An History of the Earth,* vol. 8: 194; "embarrass, v.," OED Online, December 2018, Oxford University Press, www.oed.com.ezp-prod1.hul.harvard.edu/view/Entry/60793?redirectedFrom=embarrass (accessed February 14, 2019); John Woodward, *An Attempt Towards a Natural History of the Fossils of England in a Catalogue of the English Fossils in the Collection of J. Woodward, M.D.* (London: F. Fayram, 1729), Part 1, Tome 1, 116.
52. Samuel Latham Mitchill to Catherine Mitchill, January 20, 1804, and December 6, 1803, Samuel Latham Mitchill Papers, William L. Clements Library, University of Michigan, Box 1. For a detailed account of Samuel Mitchill's pursuits in natural history, see D. Graham Burnett, *Trying Leviathan: The Nineteenth-Century New York Court Case That Put the Whale on Trial and Challenged the Order of Nature* (Princeton: Princeton University Press, 2007).
53. Samuel Latham Mitchill to Catherine Mitchill, January 20, 1804, Samuel Latham Mitchill Papers, William L. Clements Library, University of Michigan, Box 1; George Shaw, *Zoological Lectures Delivered at the Royal Institution in the Years 1806 and 1807* (London: G. Kearsley, 1809), vol. 2: 213.
54. Michele Currie Navakas, "Antebellum Coral," *American Literature* 91, no. 2 (June 2019): 263–293; Danielle Coriale, "When Zoophytes Speak: Polyps and Naturalist Fantasy in the Age of Liberalism," *Nineteenth-Century Contexts: An Interdisciplinary Journal* 34, no. 1 (2012): 19–36; James C. Scott, *Weapons of the Weak: Everyday Forms of Peasant Resistance* (New Haven: Yale University Press, 1985), 36.
55. Martyn, *A New Dictionary of Natural History,* vol. 1, preface; Smellie, *The Philosophy of Natural History,* vol. 1: 2, 4–5.
56. Hagit Kvitt et al., "Breakdown of Coral Colonial Form Under Reduced pH Conditions Is Initiated in Polyps and Mediated Through Apoptosis," *Proceedings of the National Academy of Sciences* 112 (February 2015): 2083; George D. Stanley Jr., "The Evolution of Modern Corals and Their Early History," *Earth-Science Reviews* 60 (2003): 195–225; Maoz Fine and Dan Tchernov, "Scleractinian Coral Species

Survive and Recover from Decalcification," *Science* 315, no. 5820 (March 2007): 1811; Mónica Medina et al., "Naked Corals: Skeleton Loss in Scleractinia," *Proceedings of the National Academy of Sciences* 103, no. 31 (August 2006): 9096–9100; Paul W. Sammarco, "Polyp Bail-Out: An Escape Response to Environmental Stress and a New Means of Reproduction in Corals," *Marine Ecology Progress Series* 10 (1982): 57–65.

57. Stanley, "The Evolution of Modern Corals and Their Early History," 221.

CHAPTER 2. GETTING BACK

1. "Scrapbook of Advertisements, Broadsides, Poetry, Newspaper Clippings, etc., 1745–1838 (bulk 1780–1800)," Folio 66 748 Sc43, Lewis Walpole Library, Yale University, 22v. Most estimates put the number of remaining Australian cassowaries between 1,200 and 1,500 individuals, although they are also difficult to track; see Peter Latch, "Recovery Plan for the Southern Cassowary, *Casuarius casuarius johnsonii*," Australian Government and Queensland Government Environmental Protection Agency, 2007, www.environment.gov.au/system/files/resources/79235f07-9c32-45fa-b868-eb248691e945/files/sth-cassowary.pdf (accessed September 7, 2022).
2. Tony Horwitz, *Blue Latitudes: Boldly Going Where Captain Cook Has Gone Before* (New York: Picador, 2002), 170; "The Bloomfield Track," Destination Daintree, www.destinationdaintree.com/locations/the-bloomfield-track (accessed September 7, 2022).
3. UNESCO, "Wet Tropics of Queensland," https://whc.unesco.org/en/list/486 (accessed September 7, 2022); Bill Wilkie, *The Daintree Blockade: The Battle for Australia's Tropical Rainforests* (Mossman, Queensland: Four Mile Books, 2017).
4. For a more literal reenactment of Cook's voyage, see Iain McCalman, "The Little Ship of Horrors: Reenacting Extreme History," *Criticism* 46, no. 3 (Summer 2004): 477–486.
5. Cook, August 16, 1770, *The Journals of Captain James Cook*, 378. Banks wrote a similar diatribe; Cook often borrowed language from his journal.
6. Peyssonnel, "An Account of a Manuscript Treatise Presented to the Royal Society," 448; Parsons, "A Letter from James Parsons, M.D. F.R.S. to the Rev. Mr. Birch," 511–512. On European aversions to swimming, see Dawson, *Undercurrents of Power*, 12–14.
7. Irus Braverman, *Coral Whisperers: Scientists on the Brink* (Oakland: University of California Press, 2018), 87.
8. Peter S. Vroom et al., "Algae-Dominated Reefs," *American Scientist* 94, no. 5 (September–October 2006): 430–437.
9. For a fuller critique of the unstated assumptions behind the notion of shifting baselines, see Irus Braverman, "Shifting Baselines in Coral Conservation," *Environment and Planning E: Nature and Space* 3, no. 1 (March 2020): 20–39.
10. Cook, June 13, 1770, *The Journals of Captain James Cook*, 347. For the shit turn in environmental history, see, for instance, Paul Kreitman, "Attacked by Excrement: The Political Ecology of Shit in Wartime and Postwar Tokyo," *Environmental His-*

tory 23, no. 2 (April 2018): 342–366; David L. Howell, "Fecal Matters: Prolegomenon to a History of Shit in Japan," in *Japan at Nature's Edge: The Environmental Context of a Global Power,* ed. Ian Jared Miller, Julia Adeney Thomas, and Brett L. Walker (Honolulu: University of Hawai'i Press, 2013), 137–151; Ed Melillo, "The First Green Revolution: Debt Peonage and the Making of the Nitrogen Fertilizer Trade, 1840–1930," *American Historical Review* 117, no. 4 (October 2012): 1028–1060. See also Dominique Laporte, *History of Shit,* trans. Nadia Benabid and Rodolphe el-Khoury (Cambridge: MIT Press, 2000 [orig. French publication 1978]).

11. Quoted in McCalman, "Turtle War: Captain Cook's Environmental Crisis on the Great Barrier Reef," *The Great Circle* 34, no. 2 (2012): 7.
12. Eric Deeral, quoted in wall text at the Cooktown Museum (formerly the James Cook Historical Museum), Cooktown, Queensland, Australia, dated 2001 (visited August 5, 2018). See also Iain McCalman, *The Reef: A Passionate History: The Great Barrier Reef from Captain Cook to Climate Change* (New York: Scientific American/Farrar, Straus, and Giroux, 2013), 23; McCalman, "Turtle War," 7–18.
13. Josie Olbar, quoted in "Endeavour Reef," National Museum of Australia, www.nma.gov.au/exhibitions/endeavour-voyage/endeavour-reef (accessed September 7, 2022); Gertie Deeral, Hope Vale Art and Culture Centre, www.hopevaleart.org.au/gertie-deeral (accessed September 7, 2022).
14. Sarah Kuta, "Shipworms Are Eating a Wreck That Could Be Captain Cook's 'Endeavour,'" *Smithsonian Magazine* (August 18, 2022), www.smithsonianmag.com/smart-news/shipworms-are-destroying-the-ship-believed-to-be-captain-cooks-endeavour-180980599/ (accessed September 9, 2022). On the controversy over the identification of the wreck, see Brian Amaral, "Is a Famous Shipwreck in Newport Harbor? An International Fight over the Answer Has Turned Personal," *Boston Globe,* July 6, 2022, www.bostonglobe.com/2022/07/06/magazine/is-famous-shipwreck-newport-harbor-rhode-island-expert-is-fighting-australia-over-answer/ (accessed October 10, 2022).
15. "The Bloomfield Track," Destination Daintree.
16. See especially Braverman, "Fragments of Hope: Nursing Corals Back to Life," in *Coral Whisperers,* 109–138.
17. For a sample obituary, see Rowan Jacobsen, "Obituary: Great Barrier Reef (25 Million BC–2016)," *Outside* (October 11, 2016), www.outsideonline.com/2112086/obituary-great-barrier-reef-25-million-bc-2016 (accessed September 7, 2022). I have retained "BC–2016" in the title as originally published; at the time of my writing this, the title has since been updated to replace 2016 with an ellipsis.
18. Braverman, *Coral Whisperers,* 3, 31; Christopher D. Wells and Kaitlyn J. Tonra, "Polyp Bailout and Reattachment of the Abundant Caribbean Octocoral *Eunicea flexuosa,*" *Coral Reefs* 40, no. 1 (2021): 27 and 29.
19. Parsons, *Philosophical Observations,* 204; J. E. N. Veron, "Mass Extinctions and Ocean Acidification: Biological Constraints on Geological Dilemmas," *Coral Reefs* 27 (2008): 459–472; Gal Dishon et al., "Evolutionary Traits that Enable Scleractinian Corals to Survive Mass Extinction Events," *Scientific Reports* 10, no. 3903

(2020): https://doi.org/10.1038/s41598-020-60605-2; Elizabeth Kolbert, *The Sixth Extinction: An Unnatural History* (New York: Picador, 2014); Gerardo Ceballos et al., "Accelerated Modern Human-Induced Species Losses: Entering the Sixth Mass Extinction," *Science Advances* 1, no. 5 (2015): e1400253; Stuart L. Pimm et al., "The Biodiversity of Species and Their Rates of Extinction, Distribution, and Protection," *Science* 344, no. 6187 (2014): 1246752.

CHAPTER 3. THE LOST SERPENT

1. Edward Tyson, "Vipera Caudi-sona Americana, Or the Anatomy of a Rattle-Snake, Dissected at the Repository of the Royal Society in January 1682/3," *Philosophical Transactions* 13 (1682/3): 28. My use of "afterlife" here follows Samuel J. M. M. Alberti; see *The Afterlives of Animals: A Museum Menagerie,* ed. Samuel J. M. M. Alberti (Charlottesville: University of Virginia Press, 2011).
2. George Albert Boulenger, *Catalogue of the Snakes in the British Museum (Natural History),* vol. 3 (London, 1896); John Edward Gray, *Catalogue of the Specimens of Snakes in the Collection of the British Museum* (London, 1849), 20. On Sloane, see Delbourgo, *Collecting the World.* On Shaw's self-styled "cremations," see "Report from the Select Committee on the Condition, Management and Affairs of the British Museum. . . . ," in *Selection of Reports and Papers of the House of Commons* (1836), 33: 197. On Leach and snake fumes, see Edward Edwards, *Lives of the Founders of the British Museum; with Notices of Its Chief Augmentors and Other Benefactors, 1570–1870* (London: Trübner, 1870), part 2, 576. On Sloane's botanical specimens (which, in stark contrast, have been dotingly preserved in the Sloane Herbarium), see Edwin D. Rose, "Specimens, Slips and Systems: Daniel Solander and the Classification of Nature at the World's First Public Museum, 1753–1768," *British Journal for the History of Science* 51, no. 2 (June 2018): 206.
3. *A Geographical History of Nova Scotia . . .* (London: Paul Vaillant, 1749), 100.
4. Benjamin Smith Barton, *An Account of the Most Effectual Means of Preventing the Deleterious Consequences of the Bite of the Crotalus horridus, or Rattle-Snake* (Philadelphia: R. Aitken and Son, 1792), 17.
5. Donna Haraway, *The Companion Species Manifesto: Dogs, People, and Significant Otherness* (Chicago: Prickly Paradigm Press, 2003), 3.
6. Ted Levin, *America's Snake: The Rise and Fall of the Timber Rattlesnake,* illustrated by Alexandra Westrich (Chicago: University of Chicago Press, 2016), 4 ("novel"); Thomas P. Slaughter, *The Natures of John and William Bartram* (Philadelphia: University of Pennsylvania Press, 2005 [1996]), 142. On the revolutionary iconography of rattlesnakes, see Robles, "The Rattlesnake and the Hibernaculum"; Zachary McLeod Hutchins, "Rattlesnakes in the Garden: The Fascinating Serpents of the Early, Edenic Republic," *Early American Studies* 9, no. 3 (Fall 2011): 677–715. For a limited sampling of studies of early American rattlesnakes through the years—almost all engaged with cultural representations—see David Scofield Wilson, "The Rattlesnake," in *American Wildlife in Symbol and Story,* ed. Angus K. Gillespie and Jay Mechling (Knoxville: University of Tennessee Press, 1987), 132–

154; Christoph Irmscher, "The Power of Fascination," in *The Poetics of Natural History: From John Bartram to William James* (New Brunswick: Rutgers University Press, 1999), 149–187; Philip Levy, "None but the Rattlesnakes!" in *Fellow Travelers: Indians and Europeans Contesting the Early American Trail* (Gainesville: University Press of Florida, 2007), 81–104. For a popular account of New England rattlesnakes, see Thomas Palmer, *Landscape with Reptile: Rattlesnakes in an Urban World* (Athens: University of Georgia Press, 2018 [1992]); for a two-volume work on the natural and cultural history of rattlesnakes from a herpetologist's perspective, see Laurence M. Klauber, *Rattlesnakes: Their Habits, Life Histories, and Influence on Mankind* (Berkeley: University of California Press, 1956).

7. Francis Higginson, *New-Englands Plantation. Or, a Short and True Description of the Commodities and Discommodities of That Countrey* (London: Printed by T. C. and R. C. for Michael Sparke, 1630; repr., Washington, D.C., 1835), 10 ("sup"), 5 ("nothing"), 12 ("This Countrey"); William Wood, *New Englands Prospect* . . . (London, 1639 [1634]), 38.
8. Mark Catesby, *The Natural History of Carolina, Florida and the Bahama Islands* . . . (London: Printed at the expence of the author, 1731–1743 [1729–1747]), vol. 2: 41; Paul Dudley, "An Account of the Rattlesnake. By the Honourable Paul Dudley, Esq; F.R.S.," *Philosophical Transactions* 32 (1722–1723): 292; Esther Louise Larsen, trans. and ed., "Pehr Kalm's Account of the North American Rattlesnake and the Medicines Used in the Treatment of Its Sting," *American Midland Naturalist* 57, no. 2 (April 1957): 509. John Bartram insisted that "we may justly admire the goodness of Providence in giveing this noxious Animal a Rattle in his Tail to give notice where he is . . . most that are bit is by accident, by treading on them at unawares as they lie coil'd up, or asleep." John Bartram to Peter Collinson, February 27, 17[3]7, in *The Correspondence of John Bartram, 1734–1777*, ed. Edmund Berkeley and Dorothy Smith Berkeley (Gainesville: University Press of Florida, 1992), 40.
9. Peter Collinson to John Bartram, February 3, 1735/6, in *The Correspondence of John Bartram*, 18; William Bartram, *Travels through North & South Carolina, Georgia, East & West Florida* . . . (Philadelphia: James and Johnson, 1791), 271; Kalm, "Pehr Kalm's Account of the North American Rattlesnake," 507. Thomas P. Slaughter has also shown how the Bartrams harbored an exceptional sympathy for snakes, one not solely reducible to their Quaker faith. See Slaughter, *Natures of John and William Bartram*, 132.
10. John Lawson, *A New Voyage to Carolina; Containing the Exact Description and Natural History of That Country* . . . (London, 1709), 129; Kevin Joel Berland, ed., *The Dividing Line Histories of William Byrd II of Westover* (Chapel Hill: Omohundro Institute of Early American History and Culture and University of North Carolina Press, 2013), 115, 121, 378. On Native presence in the disputed region of Byrd's survey, see Shannon Lee Dawdy, "The Meherrin's Secret History of the Dividing Line," *North Carolina Historical Review* 72, no. 4 (October 1995): 387–415; Angela Calcaterra, "Locating American Indians Along William Byrd II's Dividing Line," *Early American Literature* 46, no. 2 (2011): 233–261. For the copperplate's associa-

tion with Byrd, see Margaret Beck Pritchard and Virginia Lascara Sites, *William Byrd II and His Lost History: Engravings of the Americas* (Williamsburg, Va.: Colonial Williamsburg Foundation, 1993).

11. Peter Collinson to John Bartram, September 20 or 22, 1751, *The Correspondence of John Bartram,* 333; Thomas Jefferson to Henry Lee, June 28, 1793, Thomas Jefferson Papers, Manuscripts Division, Library of Congress, Washington, D.C.; Kalm, "Pehr Kalm's Account of the North American Rattlesnake," 506. On the ongoing scientific debate about inborn human fear of snakes, see, for instance, Edward O. Wilson, "The Serpent," in *Biophilia: The Human Bond with Other Species* (Cambridge: Harvard University Press, 2003 [1984]), 83–102; Quan Van Le et al., "Pulvinar Neurons Reveal Neurobiological Evidence of Past Selection for Rapid Detection of Snakes," *Proceedings of the National Academy of Sciences* 110, no. 47 (November 19, 2013): 19000–19005. On historicizing disgust, see Sara Ahmed, *The Cultural Politics of Emotion* (Edinburgh: Edinburgh University Press, 2004); Nancy Shoemaker, "Whale Meat in American History," *Environmental History* 10, no. 2 (April 2005): 269–294. On emotions and illness in the early modern period, see Olivia Weisser, *Ill Composed: Sickness, Gender, and Belief in Early Modern England* (New Haven: Yale University Press, 2015), 83–84.
12. All quotations from Breintnall, "A Letter from Mr. J. Breintnal to Mr. Peter Collinson, F.R.S. Containing an Account of What He Felt After Being Bit by a Rattle-Snake," *Philosophical Transactions* 44 (1746): 147–150.
13. On Breintnall's apparent suicide, see Irmscher, *The Poetics of Natural History,* 154. Few recent scientific studies have examined the mental dimension of snakebites, such as psychotic effects or longer-term mental health concerns. One exception is Shehan S. Williams et al., "Delayed Psychological Morbidity Associated with Snakebite Envenoming," *PLoS Neglected Tropical Diseases* 5, no. 8 (August 2011): e1255.
14. Benjamin Smith Barton, *A Memoir Concerning the Fascinating Faculty Which Has Been Ascribed to the Rattle-Snake, and Other American Serpents* (Philadelphia: Henry Sweitzer, 1796), 45. For a fuller account of Barton's complicated relationship with rattlesnakes in light of his fears, see Robles, "The Rattlesnake and the Hibernaculum," 36–41.
15. Goldsmith, *An History of the Earth,* vol. 7: 159–160.
16. Goldsmith, *An History of the Earth,* vol. 7: 161, 171–172; Kalm, "Sloane's Museum at Chelsea, as described by Per Kalm, 1748," trans. William R. Mead, in *Sir Hans Sloane: Collector, Scientist, Antiquary,* 33–34; Buffon, *The Natural History of Oviparous Quadrupeds and Serpents. Arranged and Published from the Papers and Collections of the Count de Buffon, by the Count De La Cepede . . . ,* trans. Robert Kerr (Edinburgh, 1802), vol. 3: 50–51.
17. Harry W. Greene et al., "Parental Behavior by Vipers," in *Biology of the Vipers,* ed. Gordon W. Schuett, Mats Höggren, Michael E. Douglas, and Harry W. Greene (Eagle Mountain, Utah: Eagle Mountain Publishing, 2002), 199. The authors reflect: "The lives of these animals are likely far more complex than even herpetologists have imagined. . . . Consider that a 'family' exists when 'offspring continue to interact into adulthood with their parents.' . . . Culture, however difficult to

define, typically implies transmission of information across generations (McGrew, 2001), so is that term applicable to initial location of dens by young-of-the-year *C. horridus* [timber rattlesnakes] . . . that follow chemical trails left by adult conspecifics?"

18. John Bartram to Peter Collinson, undated [fall of 1753], *The Correspondence of John Bartram,* 361; *The Dividing Line Histories of William Byrd II,* 125.
19. Letter from Cotton Mather to Richard Waller, November 27, 1712, EL/M2/31, Royal Society Archives. On community rattlesnake hunts, see, for instance, *History of the Town of Dorchester, Massachusetts, by a Committee of the Dorchester Antiquarian and Historical Society* (Boston: Ebenezer Clapp Jr., 1859), 351; Levin, *America's Snake,* xx. For a history of bounties offered in early America, particularly for wolves, see Jon T. Coleman, *Vicious: Wolves and Men in America* (New Haven: Yale University Press, 2004).
20. Dudley, "Account of the Rattlesnake," 295; Kalm, "Pehr Kalm's Account of the North American Rattlesnake," 504; William S. Brown, Len Jones, and Randy Stechert, "A Case in Herpetological Conservation: Notorious Poacher Convicted of Illegal Trafficking in Timber Rattlesnakes," *Bulletin of the Chicago Herpetological Society* 29, no. 4 (April 1994): 74–79; John Bartram to Michael Collinson, November 11, 1772, in Berkeley and Berkeley, *Correspondence,* 752.
21. An American Guesser, "Messrs. Printers," *Pennsylvania Journal; and the Weekly Advertiser,* December 27, 1775, [1]; Rulon W. Clark, "Kin Recognition in Rattlesnakes," *Proceedings of the Royal Society B: Biological Sciences* 271, suppl. 4 (May 2004): S243–S245 (quotation, S243); Rulon W. Clark et al., "Cryptic Sociality in Rattlesnakes (*Crotalus horridus*) Detected by Kinship Analysis," *Biology Letters* 8, no. 4 (August 2012): 523–525.
22. William S. Brown, "Female Reproductive Ecology in a Northern Population of the Timber Rattlesnake, *Crotalus horridus,*" *Herpetologica* 47, no. 1 (March 1991): 101–115.
23. Edward Bancroft, *An Essay on the Natural History of Guiana, in South America . . .* (London: T. Becket and P. A. De Hondt, 1769), 120; "Minutes of a Meeting of the Society," Journal Book of the Royal Society, JBO/10/39, July 20, 1697, Royal Society Archives; "Minutes of a Meeting of the Society," Journal Book of the Royal Society, JBO/10/133, February 7, 1699/1700, Royal Society Archives.
24. *London Daily Post and General Advertiser* (London, England), Saturday, April 29, 1738, Issue 1092, Gale Database Seventeenth and Eighteenth Century Burney Newspapers Collection; Caroline Grigson, *Menagerie: The History of Exotic Animals in England* (Oxford: Oxford University Press, 2016), 55 ("extremely"); *London Daily Post and General Advertiser* (London, England), Saturday, March 3, 1739, Issue 1356, Gale Database Seventeenth and Eighteenth Century Burney Newspapers Collection. On early modern science's fascination with reproductive organs, sex, and gender, see Londa Schiebinger, *Nature's Body: Gender in the Making of Modern Science* (New Brunswick: Rutgers University Press, 2013 [1993]); Susan Scott Parrish, "The Female Opossum and the Nature of the New World," *William and Mary Quarterly* 54, no. 3 (July 1997): 475–514; Katharine Park, *Secrets of Women:*

Gender, Generation, and the Origins of Human Dissection (New York: Zone Books, 2006).

25. George Hamell and William A. Fox, "Rattlesnake Tales," *Ontario Archaeology,* nos. 79/80 (2005): 127–149.
26. For additional information on Lower, Middle, and Upper Worlds in the phenomenologies of southeastern Native peoples, see Jodi A. Byrd, *The Transit of Empire: Indigenous Critiques of Colonialism* (Minneapolis: University of Minnesota Press, 2011), xxvii; Daniel Heath Justice, *Our Fire Survives the Storm: A Cherokee Literary History* (Minneapolis: University of Minnesota Press, 2006), 27–28; LeAnne Howe, "The Chaos of Angels," *Callaloo* 17, no. 1 (Winter 1994): 108–114.
27. Blanket, quoted in Cephas Washburn, *Reminiscences of the Indians* (Richmond, Va.: Presbyterian Committee of Publication, 1869), 208; Robert J. Conley, *A Cherokee Encyclopedia* (Albuquerque: University of New Mexico Press, 2007), 30.
28. Blanket, quoted in *Reminiscences of the Indians,* 209–211.
29. James Adair, *The History of the American Indians . . .* (London: Edward and Charles Dilly, 1775), 238; Kalm, "Pehr Kalm's Account of the North American Rattlesnake," 509; John Heckewelder to Benjamin Smith Barton, September 26, 1795, Violetta W. Delafield Collection of the Papers of Benjamin Smith Barton (Mss.B.B284d), American Philosophical Society, Philadelphia.
30. Adair, *History of the American Indians,* 238–239. For another period account of defanging, see John Brickell, *The Natural History of North-Carolina . . .* (Dublin: James Carson, 1737), 145. On the "ecological Indian" trope, see Shepard Krech III, *The Ecological Indian: Myth and History* (New York: W. W. Norton, 1999). Krech's book generated substantial controversy, particularly for how it has been deployed to deny Native sovereignty. See Vine Deloria Jr., "The Speculations of Krech," review of *The Ecological Indian,* by Shepard Krech, *Worldviews* 4, no. 3 (2000): 283–293; Michael E. Harkin and David Rich Lewis, eds., *Native Americans and the Environment: Perspectives on the Ecological Indian* (Lincoln: University of Nebraska Press, 2007). Native scholars have shown that, in broad strokes, Indigenous cosmologies stress mutual obligations between human and nonhuman beings to a greater extent than Western science. See for instance Megan Bang, Ananda Marin, and Douglas Medin, "If Indigenous Peoples Stand with the Sciences, Will Scientists Stand with Us?" *Daedalus* 147, no. 2 (Spring 2018): 148–159.
31. Bartram, *Travels,* 261.
32. Bartram, *Travels,* 271.
33. Thomas Jefferson, *Notes on the State of Virginia; With an Appendix,* third American edition (New York: M. L. & W. A. Davis, 1801), 77. On how rapid changes in colonial island environments ushered an understanding of the malleability of nature, see Richard H. Grove, *Green Imperialism: Colonial Expansion, Tropical Island Edens, and the Origins of Environmentalism, 1600–1860* (Cambridge: Cambridge University Press, 1995). On how people came to understand humanity's role in extinction, see Mark V. Barrow Jr., *Nature's Ghosts: Confronting Extinction from the Age of Jefferson to the Age of Ecology* (Chicago: University of Chicago Press, 2009).

34. Michael Collinson to John Bartram, January 8, 1773, in Berkeley and Berkeley, *Correspondence,* 755. On studies of extinction outside of natural philosophy that even predate the eighteenth century, see Vera Keller, "Nero and the Last Stalk of *Silphion:* Collecting Extinct Nature in Early Modern Europe," *Early Science and Medicine* 19, no. 5 (2014): 424–447.
35. John Bartram to Michael Collinson, November 11, 1772, in Berkeley and Berkeley, *Correspondence,* 752.
36. John Bartram to Michael Collinson, November 11, 1772, in Berkeley and Berkeley, *Correspondence,* 752.
37. David McClure, *Diary of David McClure, Doctor of Divinity: 1748–1820,* notes by Franklin B. Dexter (New York: Knickerbocker Press, 1899), 125 ("dazzling"); Elias Willard, quoted in Samuel Williams, *The Natural and Civil History of Vermont* (Burlington, Vt.: Samuel Mills, 1809), 2nd ed., vol. 1: 487 ("rapturous"); Bernardino de Sahagún, *Book 11: Earthly Things,* pt. 12 of *Florentine Codex: General History of the Things of New Spain,* ed. and trans. Arthur J. O. Anderson and Charles E. Dibble (Santa Fe, N. Mex.: Published by the School of American Research and the University of Utah, 2012 [1963]), 75 ("rainbow").
38. Sir Hans Sloane, "Conjectures on the Charming or Fascinating Power Attributed to the Rattle-Snake: Grounded on Credible Accounts, Experiments and Observations," *Philosophical Transactions* 38 (1734): 321–331, esp. 323; William Byrd, quoted in Peter Collinson to John Bartram, December 10, 1737, in Berkeley and Berkeley, *Correspondence,* 68.
39. Sue Ann Prince, introduction to *Stuffing Birds, Pressing Plants, Shaping Knowledge: Natural History in North America, 1730–1860,* ed. Sue Ann Prince (Philadelphia: American Philosophical Society, 2003), 1–9 ("outdated," 5); Levin, *America's Snake,* 12 ("not true"); Herbert Leventhal, "The Fascinating Rattlesnake," in *In the Shadow of the Enlightenment: Occultism and Renaissance Science in Eighteenth-Century America* (New York: New York University Press, 1976), 146 ("needless"). For a fuller discussion of snake enchantment and possible explanations, see Robles, "The Rattlesnake and the Hibernaculum."
40. Goldsmith, *An History of the Earth,* vol. 7: 160; Gayatri Chakravorty Spivak, "Can the Subaltern Speak?" in *Marxism and the Interpretation of Culture,* ed. Cary Nelson and Lawrence Grossberg (Urbana: University of Illinois Press, 1988), 271–313.
41. Manuscript Catalogue of Sir Hans Sloane's Collections, vol. 1, Serpents, Specimen 307, p. 147, Natural History Museum (London).
42. Manuscript Catalogue of Sir Hans Sloane's Collections, vol. 1, Serpents, Specimen 130, p. 122; Sloane, "Conjectures on the Charming or Fascinating Power Attributed to the Rattle-Snake," 322.
43. Some kill rattlesnakes due to a perceived public duty to protect people; some are motivated by the commercial pet trade; and some capture rattlesnakes for sport or as part of communal rattlesnake roundups. See William S. Brown, *Biology, Status, and Management of the Timber Rattlesnake (*Crotalus Horridus*): A Guide for Conservation,* Society for the Study of Amphibians and Reptiles Herpetological Circulars no. 22 (1993), 28–30, 44–45 ("essential"). A conservation plan from the Massachu-

setts Division of Fisheries and Wildlife likewise insists that "maintaining a level of secrecy about their locations is important for restricting human access to key habitat features in order to avoid disturbing and stressing snakes. Additionally, there continues to be a need to limit and eliminate trails on public lands near dens and basking areas." "Timber Rattlesnake, *Crotalus horridus,*" Natural Heritage and Endangered Species Program, p. 3, www.mass.gov/files/documents/2016/08/no/crotalus-horridus.pdf (accessed September 9, 2022).

44. See C. L. Fincher and D. Duvall, "Intense Hunting by Humans May Reduce Rattling in Rattlesnakes [Abstract]," *American Zoologist* 38, no. 5 (1998): 170A; Garry Ellenbolt, "Some Rattlesnakes Losing Their Warning Rattle in S. Dakota," August 29, 2013, *All Things Considered,* National Public Radio, www.npr.org/templates/story/story.php?storyId=216924322 (accessed September 9, 2022). For an article that weighs the evidence, see Jaclyn Moyer, "Don't Tread on Me," *Orion Magazine,* April 11, 2019, https://orionmagazine.org/article/dont-tread-on-me/ (accessed September 9, 2022).

CHAPTER 4. SECRETS

1. As you may have guessed by now, I also can't cite the blacklisted book here.
2. Brown, Jones, and Stechert, "Case in Herpetological Conservation," 74–75, 78.
3. Bartram, *Travels,* 264.
4. Each time Guillermo has told this story, the precise wording has varied slightly in both languages: There is nothing more poisonous, venomous, potent, dangerous, powerful—in Spanish, *venenoso, peligroso, potente, poderoso*—than a pregnant woman.
5. Kalm, "Pehr Kalm's Account of the North American Rattlesnake," 507 ("wide"), 509 ("easy").
6. For a history of wildlife tracking technologies, see Etienne Benson, *Wired Wilderness: Technologies of Tracking and the Making of Modern Wildlife* (Baltimore: Johns Hopkins University Press, 2010).
7. Deirdre Cooper Owens, *Medical Bondage: Race, Gender, and the Origins of American Gynecology* (Athens: University of Georgia Press, 2017); Deirdre Cooper Owens and Sharla M. Fett, "Black Maternal and Infant Health: Historical Legacies of Slavery," *American Journal of Public Health* 109, no. 10 (2019): 1342–1345; Nora Doyle, *Maternal Bodies: Redefining Motherhood in Early America* (Chapel Hill: University of North Carolina Press, 2018), 42–43; Terri Kapsalis, "Mastering the Female Pelvis: Race and the Tools of Reproduction," in *Skin Deep, Spirit Strong: The Black Female Body in American Culture,* ed. Kimberly Wallace-Sanders (Ann Arbor: University of Michigan Press, 2002), 263–300; Marie Jenkins Schwartz, *Birthing a Slave: Motherhood and Medicine in the Antebellum South* (Cambridge: Harvard University Press, 2010). On the broader violence of pregnancy under enslavement, see, for example, Jennifer L. Morgan, *Reckoning with Slavery: Gender, Kinship, and Capitalism in the Early Black Atlantic* (Durham, N.C.: Duke University Press, 2021), esp. 155–156 and 165–167.
8. J[onathan] Carver, *Travels through the Interior Parts of North America, in the Years 1766, 1767, and 1768* (London, 1778), 485.

9. Levin, *America's Snake,* 230.
10. Sarah Knott, *Mother Is a Verb: An Unconventional History* (New York: Farrar, Straus, and Giroux, 2019), 60; Brown, Jones, and Stechert, "Case in Herpetological Conservation," 75. For one example of the "greater catchability" of gravid snakes, see W. H. Martin, "Reproduction of the Timber Rattlesnake (*Crotalus horridus*) in the Appalachian Mountains," *Journal of Herpetology* 27, no. 2 (June 1993): 133–143.
11. Mark A. Norell et al., "A Nesting Dinosaur," *Nature* 378 (December 1995): 774–776; Ed Yong, "Dinosaur Daddies Took Care of Their Young Alone," *National Geographic* (December 18, 2008), www.nationalgeographic.com/science/phenomena/2008/12/18/dinosaur-daddies-took-care-of-their-young-alone/ (accessed June 12, 2020). For one fascinating early recognition (in 2002) of how widespread parenting is among vipers like rattlesnakes, see Greene et al., "Parental Behavior by Vipers."
12. Willard, quoted in Williams, *The Natural and Civil History of Vermont,* vol. 1: 487.
13. Robert Macfarlane, *Underland: A Deep Time Journey* (New York: W. W. Norton, 2019), 8; Brown, Jones, and Stechert, "Case in Herpetological Conservation," 74–75; William Cronon, "The Trouble with Wilderness; or, Getting Back to the Wrong Nature," in *Uncommon Ground: Rethinking the Human Place in Nature,* ed. William Cronon (New York: W. W. Norton, 1995), 79; William M. Denevan, "The Pristine Myth: The Landscape of the Americas in 1492," *Annals of the Association of American Geographers* 82, no. 3 (September 1992): 369–385.

CHAPTER 5. THE FLOUNDER AND THE RAY

1. On scientific "paperwork" in the early modern period, see Staffan Müller-Wille and Isabelle Charmantier, "Natural History and Information Overload: The Case of Linnaeus," *Studies in History and Philosophy of Science Part C: Studies in History and Philosophy of Biological and Biomedical Sciences* 43, no. 1 (March 2012): 4–15; Matthew C. Hunter, "Knives Out: Thinking On, With, Through, and Against Paper in the Mid-1660s," in *Wicked Intelligence: Visual Art and the Science of Experiment in Restoration London* (Chicago: University of Chicago Press, 2013). See also Peter Harrison, "Linnaeus as a Second Adam? Taxonomy and the Religious Vocation," *Zygon* 44, no. 4 (December 2009): 879–893.
2. Bruno Latour, "Visualisation and Cognition: Drawing Things Together," *Knowledge and Society: Studies in the Sociology of Culture Past and Present,* ed. H. Kuklick 6 (1986), 15. For more on the "visual epistemology" of natural history, see Daniela Bleichmar, *Visible Empire: Botanical Expeditions & Visual Culture in the Hispanic Enlightenment* (Chicago: University of Chicago Press, 2012), 8. On the rise of images in early modern science and medicine, see Sachiko Kusukawa, *Picturing the Book of Nature: Image, Text, and Argument in Sixteenth-Century Human Anatomy and Medical Botany* (Chicago: University of Chicago Press, 2012). For additional scholarship concerned with flatness and flattening, see B. W. Higman, *Flatness* (London: Reaktion Books, 2017); David Summers, *Real Spaces: World Art History and the Rise of Western Modernism* (London: Phaidon, 2003); Nick Sousanis, *Unflattening* (Cambridge: Harvard University Press, 2015); Whitney Barlow Robles,

"Flatness," in *The Philosophy Chamber: Art and Science in Harvard's Teaching Cabinet, 1766–1820,* ed. Ethan W. Lasser (New Haven and Cambridge: Yale University Press and Harvard Art Museums, 2017), 190–209.

3. Latour, "Visualisation and Cognition," 15; Mark Catesby, *The Natural History of Carolina, Florida and the Bahama Islands,* vol. 1: xi. See also Bruno Latour, "Centres of Calculation," in *Science in Action: How to Follow Scientists and Engineers Through Society* (Cambridge: Harvard University Press, 1987).
4. Christopher Pinney, "Things Happen: Or, From Which Moment Does That Object Come?" in *Materiality,* ed. Daniel Miller (Durham, N.C.: Duke University Press, 2005), 268.
5. See also Lulu Miller, *Why Fish Don't Exist: A Story of Loss, Love, and the Hidden Order of Life* (New York: Simon and Schuster, 2020), esp. 170–175.
6. Letter from William Dandridge Peck to the Reverend Jeremy Belknap, March 30, 1790, Papers of William Dandridge Peck, Harvard University Archives, HUG 1677, Box 1; Benjamin Smith Barton, *A Discourse on Some of the Principal Desiderata in Natural History, and on the Best Means of Promoting the Study of this Science in the United States* (Philadelphia: Denham and Town, 1807), 27, 31–32. On early national natural history, see Kariann Akemi Yokota, *Unbecoming British: How Revolutionary America Became a Postcolonial Nation* (Oxford: Oxford University Press, 2011); Anya Zilberstein, "Making and Unmaking Local Knowledge in Greater New England," *Journal for Eighteenth-Century Studies* 36, no. 4 (December 2013): 559–569.
7. Alexander Garden to Carl Linnaeus, January 2, 1760, in *A Selection of the Correspondence of Linnaeus,* vol. 1: 300; William Swainson, *Taxidermy; with the Biography of Zoologists, and Notices of Their Works* (London: Longman, Orme, Brown, Green, and Longmans, 1840), 52; Samuel Latham Mitchill, "Report on Zoology," 1817, David Bates Douglass Papers, William L. Clements Library, The University of Michigan, Box 1.
8. William Dandridge Peck, Ichthyology Lecture Notes, Papers of William Dandridge Peck, HUG 1677, Box 1, Harvard University Archives. On the *Historia Piscium,* see Sachiko Kusukawa, "The *Historia Piscium* (1686)," *Notes and Records of the Royal Society of London* 54, no. 2 (2000): 179–197.
9. Peck to Benjamin Smith Barton, October 6, 1794, Violetta W. Delafield Collection of the Papers of Benjamin Smith Barton (Mss.B.B284d), American Philosophical Society ("Put me right"); Peck to James Edward Smith, September 30, 1807, Linnean Collections, GB-110/JES/COR/8/14, Linnean Society of London ("Did you ever"). Years after Peck's grand tour of Europe, the English botanist William Jackson Hooker said that Peck relayed this origin story to him; see William Jackson Hooker, "On the Botany of America," *Edinburgh Journal of Science* 2, no. 3 (1825): 118. Peck's personal annotated copy of Linnaeus's work, held in the University of Southern California's Hancock Natural History Collection, does not appear to show signs of water damage.
10. Alexander Garden to Linnaeus, January 2, 1760, in *A Selection of the Correspondence of Linnaeus,* vol. 1: 301 ("finny tribe"); Peck to William Bentley, March 2, 1815, Papers of William Bentley, HUG 1203.5, Box 1, Harvard University Archives

("form and disposition"); William D. Peck, "Description of Four Remarkable Fishes; Taken near the Piscataqua in New Hampshire," *Memoirs of the American Academy of Arts and Sciences* 2, no. 2 (1804): 46–57. (Peck wrote the *Memoirs* article in 1794, but it was not printed until 1804.) On the primacy of the outwardly visible versus the internal in eighteenth-century natural history, see also Burnett, *Trying Leviathan,* 11; Foucault, "Classifying," in *The Order of Things,* 125–165.

11. Jan Frederik Gronovius, quoted in Helen A. Choate, "An Unpublished Letter by Gronovius," *Torreya* 16, no. 5 (May 1916): 119 ("invented"); John Frid. Gronovius, "A Method of Preparing Specimens of Fish, by Drying Their Skins, as Practised by John Frid. Gronovius, M.D. at Leyden," *Philosophical Transactions* 42 (1742): 57–58. Given that Peck was reading the *Philosophical Transactions* by the time he prepared his fish specimens in the 1790s and given his clear familiarity with Gronovius's work as expressed in his lecture manuscripts, I believe Peck most likely consulted these instructions to prepare his own fish specimens—and Gronovius's recipe was certainly the most widely circulated set of such instructions.
12. Garden to Linnaeus, January 2, 1760, in *A Selection of the Correspondence of Linnaeus,* vol. 1: 300; John Bartram to Gronovius, December 6, 1745, in *Memorials of John Bartram and Humphry Marshall,* ed. William Darlington (Philadelphia: Lindsay and Blakiston, 1849), 352. See also Alwyne C. Wheeler, "The Gronovius Fish Collection: A Catalogue and Historical Account," *Bulletin of the British Museum (Natural History)* 1, no. 5 (1958): 200.
13. Peter Davis writes that "the fish specimens which have survived the ravages of time better than any other are the skins, preserved simply by sun-drying or with the use of salt." See Peter Davis, "Collecting and Preserving Fishes: A Historical Perspective," in *Naturalists in the Field: Collecting, Recording and Preserving the Natural World from the Fifteenth to the Twenty-First Century,* ed. Arthur MacGregor (Leiden: Brill, 2018), 160. On type specimens, see Lorraine Daston, "Type Specimens and Scientific Memory," *Critical Inquiry* 31, no. 1 (Autumn 2004): 153–182.
14. Swainson, *Taxidermy,* 52. See also Robert McCracken Peck, "Alcohol and Arsenic, Pepper and Pitch: Brief Histories of Preservation Techniques," in *Stuffing Birds, Pressing Plants, Shaping Knowledge,* 36–38.
15. George Humphrey, "Directions for Collecting and Preserving All Kinds of Natural Curiosities," 15; Janice Neri, *The Insect and the Image: Visualizing Nature in Early Modern Europe, 1500–1700* (Minneapolis: University of Minnesota Press, 2011), xii.
16. Linnaeus, quoted in Peck, "Alcohol and Arsenic, Pepper and Pitch," 37–38 ("in his cupboards"); Albert C. L. G. Günther, "The President's Anniversary Address," *Proceedings of the Linnean Society of London* 111 (1898–1899): 16 ("like specimens"); Gronovius in Choate, "An Unpublished Letter by Gronovius," 118 ("dryed fishes"); Peter Collinson to John Bartram, May 28, 1766, Bartram Family Papers, Historical Society of Pennsylvania, Collection 36, Box 3 ("I am pleased"). Linnaeus chose not to bind his herbarium because doing so would have made rearranging species more difficult; instead, he used a large cabinet that allowed for easy swapping of sheets, approaching his paper specimens more as files than as book leaves. See

Staffan Müller-Wille, "Linnaeus' Herbarium Cabinet: A Piece of Furniture and Its Function," *Endeavour* 30, no. 2 (2006): 60–64.

17. Thomas Browne, in *Sir Thomas Browne: The Major Works,* ed. C. A. Patrides (London: Penguin, 1977), 78–79. An unauthorized version of Browne's *Religio Medici* was released in 1642, followed by an authorized version in 1643. See also *The Book of Nature in Early Modern and Modern History,* ed. Klaas van Berkel and Arjo Vanderjagt (Leuven: Peeters, 2006).
18. John Thornton Kirkland, quoted in Obituary Notice of Professor Peck, Papers of William Dandridge Peck, Harvard University Archives, HUG 1677; William Dandridge Peck, Lecture Notes, Botany Lecture 11, Papers of William Dandridge Peck, Harvard University Archives, HUG 1677, Box 3. On the Philosophy Chamber, see Robles, "Flatness," in *The Philosophy Chamber.*
19. William Dandridge Peck, Ichthyology Lecture Notes, Papers of William Dandridge Peck, Harvard University Archives, HUG 1677, Box 1.
20. Gronovius in Choate, "An Unpublished Letter by Gronovius," 118; Peck, "Description of Four Remarkable Fishes," 55–56; John Reinhold Forster, "An Account of Some Curious Fishes, Sent from Hudson's Bay; By Mr. John Reinhold Forster, F.R.S. in a Letter to Thomas Pennant, Esq.; F.R.S.," *Philosophical Transactions* 63 (1773): 149–160.
21. College Books, 1636–1827, Harvard University, College Book 8, 1778–1803, UAI 5.5, Box 6, page 12, Harvard University Archives ("Miss Meriam").
22. Francis Goelet, entry for October 25, 1750, *The Voyages and Travels of Francis Goelet, 1746–1758,* ed. Kenneth Scott (New York: Queens College Press/Gregg Press, 1970).
23. See also Christine DeLucia, "Fugitive Collections in New England Indian Country: Indigenous Material Culture and Early American History Making at Ezra Stiles's Yale Museum," *William and Mary Quarterly* 75, no. 1 (2018): 109–150; Craig Steven Wilder, *Ebony & Ivy: Race, Slavery, and the Troubled History of America's Universities* (New York: Bloomsbury Press, 2013), esp. 199. Wilder, in his damning study of the entanglement of American universities with slavery, describes how the personal physician of Dartmouth College's first president used the skin of a deceased Black man, Cato, to cover his instrument case (at a time when the college accumulated the bones of African-descended people, both enslaved and free, from the local community for its medical collection).
24. On the notion of "fugitive science," see Britt Rusert, *Fugitive Science: Empiricism and Freedom in Early African American Culture* (New York: New York University Press, 2017), esp. 4–5.
25. Alexander Garden to Linnaeus, June 20, 1771, in *A Selection of the Correspondence of Linnaeus,* vol. 1: 331. The Providence Island mentioned here is most likely New Providence in the Bahamas rather than Old Providence or Isla de Providencia, located off the coast of Nicaragua, given Garden's nod to "the Bahama islands" and their proximity to Garden. (Other regional eighteenth-century writers, such as Catesby, also described New Providence as Providence.) On Garden's ties to slavery and Charleston as a major hub of the slave trade, see Edmund Berkeley and Dorothy Smith Berkeley, *Dr. Alexander Garden of Charles Town* (Chapel Hill:

University of North Carolina Press, 1969); Gregory E. O'Malley, "Slavery's Converging Ground: Charleston's Slave Trade as the Black Heart of the Lowcountry," *William and Mary Quarterly* 74, no. 2 (April 2017): 271–302.

26. Although his labor was effaced in the eighteenth century, this man would be celebrated in the nineteenth, even if it was still in racialized terms that treated a skilled African man as an anomaly. An entry devoted to Garden's life in an 1833 issue of *The New-York Mirror* states that Garden "had a black servant, who was peculiarly dexterous at this operation, so as to preserve the outline of the fish, and every important character, with great accuracy. . . . In the summer of 1770, Dr. Garden sent an intelligent negro servant to the island of Providence, for the purpose of collecting natural curiosities, and especially fish; this being, as we presume, the man abovementioned who excelled in their preparation." See T., "Brief Notices of Eminent Persons: Alexander Garden," *New-York Mirror: A Weekly Journal, Devoted to Literature and the Fine Arts* 10, no. 41 (April 13, 1833): 325. But it seems knowledge of his labor was lost again: my contacts at the Linnean Society had not heard of his contribution to the collection when I initially reached out for more information.
27. See also Steven Shapin, "The Invisible Technician," *American Scientist* 77 (1989): 554–563.
28. Peck, "Description of Four Remarkable Fishes," 46.
29. Peck to William Bentley, March 2, 1815, Papers of William Bentley, HUG 1203.5, Box 1, Harvard University Archives ("by a Piscataqua fisherman"); Peck, "Description of Four Remarkable Fishes," 46 ("by a boy"). See also Murphy, "Translating the Vernacular"; Marcy Norton, "Subaltern Technologies and Early Modernity in the Atlantic World," *Colonial Latin American Review* 26, no. 1 (2017): 18–38.
30. Garden to Linnaeus, June 2, 1763, in *A Selection of the Correspondence of Linnaeus*, vol. 1: 313. On the specimen and its lectotype status (a retroactive type designation), see Richard S. McBride et al., "A New Species of Ladyfish, of the Genus *Elops* (Elopiformes: Elopidae), from the Western Atlantic Ocean," *Zootaxa* 2346, no. 1 (January 2010): 29–41. For other discussions of this ladyfish specimen, see Whitney Barlow Robles, "Natural History in Two Dimensions," *Commonplace: The Journal of Early American Life* 18, no. 1 (Winter 2018), http://commonplace.online/article/vol-18-no-1-robles/ (accessed April 15, 2019); Whitney Barlow Robles, "Ladyfish," *The Kitchen in the Cabinet: Histories of Food and Science* (2021), https://kitcheninthecabinet.com/ladyfish/ (accessed January 25, 2022).
31. Daston, "Type Specimens and Scientific Memory," 158; Sir Charles Linné, *A General System of Nature . . .* , trans. William Turton (London: Lackington, Allen, and Co., 1802), vol. 1: 862. For a social and cultural history of enslaved cooks in Virginia, see Kelley Fanto Deetz, *Bound to the Fire: How Virginia's Enslaved Cooks Helped Invent American Cuisine* (Lexington: University Press of Kentucky, 2017). Deetz suggests slaveholders preferred enslaved men to women as cooks until the end of the eighteenth century, at which point women finally started to dominate kitchen spaces (see especially 130–131). Emma C. Spary, albeit with regional differences, suggests male cooks were used as markers of wealth rather than actually serving as the majority of cooks. See E. C. Spary, *Eating the Enlightenment: Food and*

the Sciences in Paris, 1670–1760 (Chicago: University of Chicago Press, 2014), 197. See also Troy Bickham, "Eating the Empire: Intersections of Food, Cookery, and Imperialism in Eighteenth-Century Britain," *Past and Present* 198, no. 1 (2008): esp. 74. For additional works showing the role of women in eighteenth-century cooking, see Rebecca Sharpless, *Cooking in Other Women's Kitchens: Domestic Workers in the South, 1865–1960* (Chapel Hill: University of North Carolina Press, 2010), 3; Alexandra Finley, "'Cash to Corinna': Domestic Labor and Sexual Economy in the 'Fancy Trade,'" *Journal of American History* 104, no. 2 (September 2017): 410–430. Joyce E. Chaplin also argues that "private life and female domestic activities, including cooking, were defined as nonessential" in earlier, exceptionalist strands of colonial American historiography, leading to the erasure of both women and food from narratives of America's founding. See Chaplin, "Food and the Material Origins of Early America," *Food in Time and Place: The American Historical Association Companion to Food History*, ed. Paul Freedman, Joyce E. Chaplin, and Ken Albala (Berkeley: University of California Press, 2014), 144. On the gender politics of medical cookery in particular, see Londa Schiebinger, "Women's Traditions," in *The Mind Has No Sex? Women in the Origins of Modern Science* (Cambridge: Harvard University Press, 1991), 112–116. Schiebinger suggests a post-1770 fracture between women who performed domestic cooking duties on one hand and men who cooked professionally, as at court, on the other.

32. Susan Leigh Star and James R. Griesemer, "Institutional Ecology, 'Translations,' and Boundary Objects: Amateurs and Professionals in Berkeley's Museum of Vertebrate Zoology, 1907–39," *Social Studies of Science* 19, no. 3 (August 1989): 387–420.
33. Garden to Linnaeus, June 2, 1763, in *A Selection of the Correspondence of Linnaeus*, vol. 1: 312–313. See also Whitney Barlow Robles, "Squid: Natural History as Food History, c. 1730–1860," in *Natural Things in Early Modern Worlds*, ed. Mackenzie Cooley, Anna Toledano, and Duygu Yıldırım (New York: Routledge, 2023).
34. Peck to Benjamin Smith Barton, March 12, 1796, Violetta W. Delafield Collection of the Papers of Benjamin Smith Barton (Mss.B.B284d), American Philosophical Society ("herbarium pests"); "Methods Selected from Various Authors, by Mr. Peck, of Preserving Animals and their Skins," *Collections of the Massachusetts Historical Society* (1795): 10–11 ("nocturnal"); Sarah Bowdich Lee, *Taxidermy: or, the Art of Collecting, Preparing, and Mounting Objects of Natural History . . .* (London: Longman, Brown, Green, and Longmans, 1843 [1820]), 204; [Samuel Kettell], *Manual of the Practical Naturalist, or, Directions for Collecting, Preparing, and Preserving Subjects of Natural History . . .* (Boston: Lilly and Wait and Carter, Hendee and Babcock, 1831), 143 ("beat"). On Sarah Bowdich Lee's scientific work, see Mary Orr, "Women Peers in the Scientific Realm: Sarah Bowdich (Lee)'s Expert Collaborations with Georges Cuvier, 1825–33," *Notes and Records* 69, no. 1 (March 2015): 37–51.
35. Letter from William Dandridge Peck to Jeremy Belknap, May 20, 1796, Jeremy Belknap Papers, Massachusetts Historical Society P-380, Reel 6.
36. Humphrey, "Directions for Collecting and Preserving All Kinds of Natural Curi-

osities," 14; Patrik Svensson, *The Book of Eels: Our Enduring Fascination with the Most Mysterious Creature in the Natural World,* trans. Agnes Broomé (New York: HarperCollins, 2021 [2019]), 24.

37. Catesby, *The Natural History of Carolina, Florida and the Bahama Islands,* vol. 2: 26. For one Anglophone recounting of Indigenous partnerships with the remora—with similar reports stretching back to at least the sixteenth century (as in the work of Spanish historian Gonzalo Fernández de Oviedo y Valdés) and resurfacing into the nineteenth—see Hans Sloane, *A Voyage to the Islands Madera, Barbados, Nieves, S. Christophers and Jamaica . . .* (London, 1707), vol. 1: 29.
38. William Bentley Papers, Folio vol. 2 (April 2, 1802), American Antiquarian Society.
39. Shaw, *Zoological Lectures,* vol. 2: 64–65.
40. See William Dandridge Peck, *Drawing of Ray Accompanying Ichthyology Lecture,* c. 1814, graphite and ink on paper, Ichthyology Lecture Notes, Papers of William Dandridge Peck, Harvard University Archives, HUG 1677, Box 5. Although I have seen several incomplete half-specimens of rays preserved on paper that show the animal's top surface, only once, in the Gronovius collection, have I ever seen an attempt to preserve a ray's top and bottom surfaces on paper—in this instance, something only possible due to the small size of the specimen.
41. Goldsmith, *An History of the Earth,* vol. 6: 247. For a fuller history of shagreen, see Christine Guth, "Towards a Global History of Shagreen," in *The Global Lives of Things: The Material Culture of Connections in the Early Modern World,* ed. Anne Gerritsen and Giorgio Riello (London: Routledge, 2015).
42. Robert Hooke, *Micrographia: Or Some Physiological Descriptions of Minute Bodies Made by Magnifying Glasses, with Observations and Inquiries Thereupon* (London: 1665), 208; Inaugural Address of Professor William Dandridge Peck, Papers of William Dandridge Peck, Harvard University Archives, HUG 1677, Box 2; George Adams, *Micrographia Illustrata: Or The Microscope Explained . . .* (London: Printed for the author, 1771 [1746]), 17, and preface.
43. "[Meetings of 1794]," *Proceedings of the Massachusetts Historical Society* 1 (1791–1835): 76; Banks, *The Endeavour Journal of Joseph Banks,* vol. 2: 28 and 7; "chagrin, n.," OED Online, December 2021, Oxford University Press, www-oed-com.dartmouth.idm.oclc.org/view/Entry/30192?result=1&rskey=09ZdyX& (accessed September 20, 2022).
44. Goldsmith, *An History of the Earth,* vol. 6: 239; Piscator [William Hughes], *Fish, How to Choose and How to Dress* (London: Longman, Brown, Green, and Longmans, 1843), 137. Some deep-sea anglerfish are also called sea devils in modern-day parlance, but Hughes is clearly discussing a large ray in this passage.
45. See also Brian P. Copenhaver, "A Tale of Two Fishes: Magical Objects in Natural History from Antiquity Through the Scientific Revolution," *Journal of the History of Ideas* 52, no. 3 (July–September 1991): 373–398.
46. Quotations from John Walsh, "Of the Electric Property of the Torpedo. In a Letter from John Walsh, Esq; F.R.S. to Benjamin Franklin . . . ," *Philosophical Transactions* 63 (1773): 461–480.
47. Walsh, "Of the Electric Property of the Torpedo." The centrality of these animals in

the history of electricity has been extensively examined in a number of works; see Marco Piccolino, "The Taming of the Electric Ray: From a Wonderful and Dreadful 'Art' to 'Animal Electricity' and 'Electric Battery,'" in *Brain, Mind, and Medicine: Neuroscience in the 18th Century*, ed. Harry Whitaker, C. U. M. Smith, and Stanley Finger (New York: Springer, 2007), 125–143; William J. Turkel, *Spark from the Deep: How Shocking Experiments with Strongly Electric Fish Powered Scientific Discovery* (Baltimore: Johns Hopkins University Press, 2013). For an account of how electric eels in Guiana also advanced studies of electricity, see James Delbourgo, "How to Handle an Electric Eel," in *A Most Amazing Scene of Wonders: Electricity and Enlightenment in Early America* (Cambridge: Harvard University Press, 2006), 165–199.

48. Letter from John Walsh to Benjamin Franklin, July 1, 1773, Royal Society of London Archive Papers (1768–1780); William Bullock, *A Companion to the London Museum, Containing a Brief Description of Several Thousand Natural and Foreign Curiosities . . .* (London: Whittingham and Rowland, 1813), 15th ed., 68.
49. Humphrey, "Directions for Collecting and Preserving All Kinds of Natural Curiosities," 12.
50. Herman Melville, *Moby-Dick* (New York: Penguin, 1998 [1851]), 125–138.
51. Goldsmith, *An History of the Earth*, vol. 2: 21; Henry David Thoreau, *The Writings of Henry David Thoreau: Journal*, ed. Bradford Torrey (Boston: Houghton Mifflin, 1906), vol. 1: 464. On this section of *Moby-Dick*, see also Parrish, *American Curiosity*, 313; Irmscher, *The Poetics of Natural History*, 70–71. In practice, Thoreau made many plant herbarium specimens, many of which are now housed at Harvard University.
52. Parrish, *American Curiosity*, esp. 310–311; Swainson, *Taxidermy*, 53.

CHAPTER 6. SPLITTING THE LARK

1. Michael Polanyi, "Tacit Knowing," in *The Tacit Dimension* (Chicago: University of Chicago Press, 2009 [1996]), 4; Pamela H. Smith, "In the Workshop of History: Making, Writing, and Meaning," *West 86th* 19, no. 1 (Spring–Summer 2012): 4–31 (esp. 12); H. Otto Sibum, "Reworking the Mechanical Value of Heat: Instruments of Precision and Gestures of Accuracy in Early Victorian England," *Studies in History and Philosophy of Science* 26, no. 1 (1995): 73–106. See also Hasok Chang, "The Myth of the Boiling Point," *Science Progress* 91, no. 3 (2008): 219–240.
2. Simona Tarricone et al., "Wild and Farmed Sea Bass (*Dicentrarchus Labrax*): Comparison of Biometry Traits, Chemical and Fatty Acid Composition of Fillets," *Fishes* 7, no. 1 (February 2022): https://doi.org/10.3390/fishes7010045; M. Vandeputte, P.-A. Gagnaire, and F. Allal, "The European Sea Bass: A Key Marine Fish Model in the Wild and in Aquaculture," *Animal Genetics* 50, no. 3 (June 2019): 195–206; "Dicentrarchus labrax (Linnaeus 1758)," Cultured Aquatic Species Information Programme, Food and Agriculture Organization of the United Nations, www.fao.org/fishery/en/culturedspecies/dicentrarchus_labrax/en (accessed September 22, 2022).
3. For another telling of this reenactment adventure, see Robles, "Natural History in

Two Dimensions." See also Simon Werrett, *Thrifty Science: Making the Most of Materials in the History of Experiment* (Chicago: University of Chicago Press, 2019).

4. Gronovius, "A Method of Preparing Specimens of Fish," 57; Manasseh Cutler, "Doctor Cutler's Method of Preserving the Skins of Birds," *Collections of the Massachusetts Historical Society* (1795): 10.
5. Gronovius, "A Method of Preparing Specimens of Fish, by Drying Their Skins," 58.
6. John Coakley Lettsom, *The Naturalist's and Traveller's Companion: Containing Instructions for Collecting and Preserving Objects of Natural History . . .* (London: E&C Dilly, 1774 [1772]), 19.
7. Gronovius, "A Method of Preparing Specimens of Fish, by Drying Their Skins," 58.
8. "Methods Selected from Various Authors, by Mr. Peck," 10.
9. William Dandridge Peck, Ichthyology Lecture Notes, Papers of William Dandridge Peck, Harvard University Archives, HUG 1677, Box 1. Some of Peck's fish have, however, been preserved on paper and then pasted (likely after the fact) on cardstock with a hanging mechanism, revealing they could at least be retrofitted for display.
10. See also Barbara Ketcham Wheaton, "Cookbooks as Resources for Social History," *Food in Time and Place.*
11. Piscator [William Hughes], *Fish, How to Choose and How to Dress,* 129 and 155; Piscator [Hughes], *A Practical Treatise on the Choice and Cookery of Fish* (London: Longman, Brown, Green, and Longmans, 1854), 140.
12. Piscator [Hughes], *A Practical Treatise on the Choice and Cookery of Fish,* 143–144; A Lady [Maria Eliza Rundell], *A New System of Domestic Cookery, Formed Upon Principles of Economy and Adapted to the Use of Private Families throughout the United States* (New York: John Forbes, 1814), 30.
13. Swainson, *Taxidermy,* 53–54; Humphrey, "Directions for Collecting and Preserving All Kinds of Natural Curiosities," 15. See also *The Kitchen in the Cabinet* (www.kitcheninthecabinet.com/); Anita Guerrini, "A Natural History of the Kitchen," *Osiris* 35 (2020): 20–41.
14. Johanna Drucker, "Humanities Approaches to Graphical Display," *Digital Humanities Quarterly* 5, no. 1 (Winter 2011): 1, www.digitalhumanities.org/dhq/vol/5/1/000091/000091.html (accessed September 23, 2022); Mitchill, "Report on Zoology"; Swainson, *Taxidermy,* 53; Lettsom, *The Naturalist's and Traveller's Companion,* 19.
15. Lauren F. Klein, *An Archive of Taste: Race and Eating in the Early United States* (Minneapolis: University of Minnesota Press, 2020), 155.
16. Thomas Barbour, "Notes on W. D. Peck by Thomas Barbour, 1895–1937," Papers of William Dandridge Peck, Harvard University Archives, HUG 1677, Box 1, Folder 22 (see letter dated November 20, 1937); Thomas Barbour, "Barbour's Correspondence About W. D. Peck, 1928–1937," Papers of William Dandridge Peck, Harvard University Archives, HUG 1677, Box 5, Folder 4 (see the letters dated December 18, 1933, and November 27, 1934); Davis, "Collecting and Preserving Fishes," 160. On the misplacement of Peck's specimens and their rediscovery in 1929, see

David P. Wheatland, *The Apparatus of Science at Harvard, 1765–1800* (Cambridge: Collection of Historical Scientific Instruments, Harvard University, 1968), 190.

17. Didi van Trijp and Robbert Striekwold, "The Ichthyologist's Garden," *The Recipes Project: Food, Magic, Art, Science, and Medicine,* July 26, 2017, https://recipes.hypotheses.org/9798 (accessed February 5, 2019); Luis Ceríaco and Mariana Marques, "Peixes Em 'Herbário.' Uma Técnica Científico-Museológica do Século XVIII," Conference Paper, Congresso Luso-Brasileiro de História das Ciências (October 2011): 1204–1219.
18. Gronovius, "A Method of Preparing Specimens of Fish, by Drying Their Skins," 58.

CHAPTER 7. SLEIGHT OF HAND

1. On the ultimate unknowability of such perspectives, see Fudge, "What Was It Like to Be a Cow?"; Thomas Nagel, "What Is It Like to Be a Bat?" *Philosophical Review* 83, no. 4 (October 1974): 435–450; Joyce E. Chaplin, "Can the Nonhuman Speak? Breaking the Chain of Being in the Anthropocene," *Journal of the History of Ideas* 78, no. 4 (October 2017): 523–524.
2. Constance Classen, "Fingerprints: Writing About Touch," in *The Book of Touch,* ed. Constance Classen (Oxford: Berg, 2005), 2. See also Céline Carayon, "Touching on Communication: Visual and Textual Representations of Touch as Friendship in Early Colonial Encounters," in *Empire of the Senses,* esp. 35–36. Modern-day biologists and eighteenth-century observers have noted that raccoons often use their hands to the exclusion of all other senses. See, for instance, Andrew N. Iwaniuk and Ian Q. Whishaw, "How Skilled Are the Skilled Limb Movements of the Raccoon (*Procyon lotor*)?" *Behavioural Brain Research* 99 (1999): 41. On the hand-brain connection in raccoons, see Samuel I. Zeveloff, *Raccoons: A Natural History* (Washington, D.C.: Smithsonian Institution Press, 2002), 70–71; W. I. Welker, J. I. Johnson Jr., and B. H. Pubols Jr., "Some Morphological and Physiological Characteristics of the Somatic Sensory System in Raccoons," *American Zoologist* 4, no. 1 (February 1964): 75–94.
3. Martin Heidegger, *What Is Called Thinking?* trans. Fred D. Wieck and J. Glenn Gray (New York: Harper and Row, 1968), 16; Steven Shapin and Barry Barnes, "Head and Hand: Rhetorical Resources in British Pedagogical Writing, 1770–1850," *Oxford Review of Education* 2, no. 3 (1976): 235. See also Fredrik Albritton Jonsson, "Enlightened Hands: Managing Dexterity in British Medicine and Manufactures, 1760–1800," in *Body Parts: Critical Explorations in Corporeality,* ed. Christopher E. Forth and Ivan Crozier (Lanham, Md.: Lexington Books, 2005), 142–158; Lissa Roberts and Simon Schaffer, preface to *The Mindful Hand: Inquiry and Invention from the Late Renaissance to Early Industrialisation,* ed. Lissa Roberts, Simon Schaffer, and Peter Dear (Amsterdam: Royal Netherlands Academy of Arts and Sciences, 2007), esp. xiii. Elizabeth D. Harvey writes of the paradoxical status of touch within European intellectual frameworks: "Touch occupies a complex, shifting, and sometimes contradictory position in the representation of the five senses in Western culture. Sometimes depicted as 'the king of senses' it was equally

likely to be disparaged as the basest sense." See Harvey, "Introduction: The 'Sense of All Senses,'" in *Sensible Flesh: On Touch in Early Modern Culture,* ed. Elizabeth D. Harvey (Philadelphia: University of Pennsylvania Press, 2003), 1.

4. John Lawson, *A New Voyage to Carolina,* 121 ("fine"); John Josselyn in *Reading the Roots: American Nature Writing Before Walden,* ed. Michael P. Branch (Athens: University of Georgia Press, 2004), 74 ("grease"). On the Algonquian (and specifically, Powhatan) word for raccoon, see Daniel Heath Justice, *Raccoon* (London: Reaktion Books, 2021), 53–54. Early transliterations of the word included *rahaughcums* (from John Smith) and *arrahacoun.* See also William R. Gerard, "Virginia's Indian Contributions to English," *American Anthropologist* 9, no. 1 (1907): 104. Gerard notes: "The animal has the reputation of being very knowing" (105).
5. Brandi L. Simmons, Jenifer Sterling, and Jane C. Watson, "Species and Size-selective Predation by Raccoons (*Procyon lotor*) Preying on Introduced Intertidal Clams," *Canadian Journal of Zoology* 92, no. 12 (2014): 1059–1065. On animal knowledge as an important part of animal history, see also Jamie Kreiner, *Legions of Pigs in the Early Medieval West* (New Haven: Yale University Press, 2020), 3–4.
6. Banks, *The Endeavour Journal of Joseph Banks,* vol. 2: 20.
7. Thomas Pennant, *Arctic Zoology* . . . (London: Henry Hughs, 1784), vol. 1: 70.
8. James Parsons, "Some Account of the Animal Sent from the East Indies, by General Clive, to His Royal Highness the Duke of Cumberland, Which Is Now in the Tower of London . . . ," *Philosophical Transactions* 51 (1760): 648–652.
9. "A Continuation of the Account of Virginia, by Mr. Clayton," in Benjamin Baddam, *Memoirs of the Royal Society; Being a New Abridgment of the Philosophical Transactions* . . . (London: G. Smith, 1739), vol. 3: 111; John Ray, *Synopsis Methodica Animalium Quadrupedum et Serpentini Generis* . . . (London: Smith and Walford, 1693), 179; Catesby, *The Natural History of Carolina, Florida and the Bahama Islands,* vol. 1: xxiv, xxix; William Cowper and Edward Tyson, "Carigueya, Seu Marsupiale Americanum Masculum . . . ," *Philosophical Transactions* 24 (1704–1705): 1565–1566 and 1569.
10. Francis Moore, *A Voyage to Georgia. Begun in Year 1735* . . . (London: Jacob Robinson, 1744), 55; James E. De Kay, "Part I. Mammalia," in *Zoology of New-York, or the New-York Fauna* . . . (Albany: W. and A. White and J. Visscher, 1842), 26.
11. Harriet Ritvo, *The Platypus and the Mermaid, and Other Figments of the Classifying Imagination* (Cambridge: Harvard University Press, 1998), xii; Blunt, *Linnaeus,* 29.
12. Carl Linnaeus, *Systema naturae per regna tria naturae* . . . , 10th ed. (Stockholm, 1758), vol. 1: 48.
13. D. D. Rasmusson and B. G. Turnbull, "Sensory Innervation of the Raccoon Forepaw: 2. Response Properties and Classification of Slowly Adapting Fibers," *Somatosensory Research* 4, no. 1 (1986): 63–75.
14. Carl Linnaeus, "Beskrifning På et Americanskt Diur, som Hans Konglige Höghet gifvit til undersökning," *Kongl. Svenska Vetenskaps Akademiens Handlingar* 8 (1747): 277–289 (quotation on 277). I am incredibly grateful to Jens Amborg for translating Linnaeus's account for me. See also Virginia C. Holmgren, *Raccoons in Folklore, History, & Today's Backyards* (Santa Barbara, Calif.: Capra Press, 1990), 54. On

Diana the monkey, see Lisbet Koerner, *Linnaeus: Nature and Nation* (Cambridge: Harvard University Press, 1999), 88–89.

15. Linnaeus, quoted in Blunt, *Linnaeus*, trans. Wilfrid Blunt, 151–152 ("This bear," "walked just like a bear"); Linnaeus, "Beskrifning På et Americanskt Diur," trans. Jens Amborg (all other quotations).
16. On Boel's painting, see also *Animal: Exploring the Zoological World* (London: Phaidon Press, 2018), 291. For more on Louis XIV's menagerie and Boel, see Peter Sahlins, *1668: The Year of the Animal in France* (New York: Zone Books, 2017). On the attribution of the drawing to Laurentius Alstrin, see S. Savage, F.L.S., "A Recently Discovered Drawing of Linnaeus's Pet Racoon," *Proceedings Linnean Society London* 150, no. 2 (March 1938), 109–110.
17. For these names and translations, see Justice, *Raccoon*, 54 and 57; Holmgren, *Raccoons in Folklore*, 157–158; Gerard, "Virginia's Indian Contributions to English," 105. While both the Holmgren and Gerard texts have outdated elements and at times treat Native cosmologies insensitively, they provide a useful starting point for Indigenous interactions with raccoons. On raccoon-bear kinship, see Justice, *Raccoon*, 18–19.
18. Nicholas J. Reo and Laura A. Ogden, "Anishnaabe Aki: An Indigenous Perspective on the Global Threat of Invasive Species," *Sustainability Science* 13, no. 5 (September 2018): 1447; George Hamell and William A. Fox, "Rattlesnake Tales," 130–131. For more on clan structures and historical political alliances between humans and nonhumans, see Vanessa Watts, "Indigenous Place-thought & Agency Amongst Humans and Non-humans (First Woman and Sky Woman Go on a European World Tour!)," *Decolonization: Indigeneity, Education, & Society* 2, no. 1 (2013): 23.
19. Sahagún, *Book 11: Earthly Things*, pt. 12 of *Florentine Codex: General History of the Things of New Spain*, 9. For a digitization of the original, see *General History of the Things of New Spain by Fray Bernardino de Sahagún: The Florentine Codex*, Library of Congress and World Digital Library, Volume 3 (Book 11) www.loc.gov/resource/gdcwdl.wdl_10096_003/?sp=339.
20. Francisco Hernández, "Historia de los Cuadrúpedos de Nueva España," in *Tomo III, Historia Natural de la Nueva España 2*, in *Las Obras Completas de Francisco Hernández* (Mexico City: UNAM, 1956–1985). I am grateful to Mackenzie Cooley for pointing me in the direction of this source. The translations of Hernández are my own.
21. Sahagún, *Book 11: Earthly Things*, pt. 12 of *Florentine Codex*, 10; Ole Worm, *Museum Wormianum . . .* (Leiden, 1655), 319. Thomas Pennant, among other authors, later identified Worm's depiction as a raccoon; see Thomas Pennant, *History of Quadrupeds* (London: B. and J. White, 1793), 3rd ed., vol. 2: 12.
22. Commonplace book (1692–1729) of Joshua Spencer, Collection 426, Academy of Natural Sciences, 231–232; Sahagún, *Book 11: Earthly Things*, pt. 12 of *Florentine Codex*, 10. See also R. A. Donkin, "The Peccary—With Observations on the Introduction of Pigs to the New World," *Transactions of the American Philosophical Society* 75, no. 5 (1985): 62.
23. Pamela H. Smith, *The Body of the Artisan: Art and Experience in the Scientific Revo-*

lution (Chicago: University of Chicago Press, 2004), 17–18; Foucault, *The Order of Things,* 133. On the emphasis on visuality, see also Mark M. Smith, "Getting in Touch with Slavery and Freedom," *Journal of American History* 95, no. 2 (September 2008): 381–382.

24. Sir Charles Bell, *The Hand, Its Mechanism and Vital Endowments, As Evincing Design* (New York: Harper and Brothers, 1840 [1833]), 40; John Ray, *The Wisdom of God Manifested in the Works of the Creation* (London: William Innys and Richard Manby, 1735 [1691]), 10th ed., Part 2, 278 and 281. On the longer history of such debates, see Fudge, *Brutal Reasoning.*
25. Buffon, *Natural History, General and Particular,* trans. William Smellie, 3rd ed., vol. 2: 8.
26. Buffon, *Natural History, General and Particular,* trans. William Smellie, 3rd ed., vol. 2: 362 ("far superior"), 366–367 ("no intermediate," "immateriality," "thinking being"); vol. 3: 294 ("degrade"). This immaterial, rational soul differed from other souls in Aristotle's philosophy, such as the sensitive soul—shared by humans and animals—and the nutritive soul—shared among humans, animals, and plants. See also Gustavo Caponi, "The Discontinuity Between Humans and Animals in Buffon's Natural History," *História, Ciências, Saúde—Manguinhos* 24, no. 1 (2017): 59–74.
27. Buffon, *Natural History, General and Particular,* trans. William Smellie, 3rd ed., vol. 3: 49.
28. Buffon, *Natural History, General and Particular,* trans. William Smellie, 3rd ed., vol. 3: 45–49.
29. Buffon, *Natural History, General and Particular,* trans. William Smellie, 3rd ed., vol. 3: 46–47, 238 ("thought and reflection").
30. William Bartram, "William Bartram Notes on the Nature of Animals with Sketches, 1780s," Bartram Family Papers, Historical Society of Pennsylvania, Collection 36, Box 1, Folder 83; William Smellie, *The Philosophy of Natural History* (Edinburgh: Charles Elliot, 1790), 156–157 ("Though no animal"), 422 ("peculiarly severe"). See also Nathaniel Wolloch, "William Smellie and the Enlightenment Critique of Anthropocentrism," in *The Enlightenment's Animals: Changing Conceptions of Animals in the Long Eighteenth Century* (Amsterdam: Amsterdam University Press, 2019), 71–87.
31. Smellie, *The Philosophy of Natural History,* 53; Benjamin Smith Barton, *Additional Facts, Observations, and Conjectures Relative to the Generation of the Opossum of North-America. In a Letter from Professor Barton to Professor J. A. H. Reimarus, of Hamburgh* (Philadelphia: S. Merritt, 1813), 21. Historians have written at great length about the "Dispute of the New World," especially the degeneracy debate as played out between Buffon and Thomas Jefferson. See, for example, Jorge Cañizares-Esguerra, *How to Write the History of the New World: Histories, Epistemologies, and Identities in the Eighteenth-Century Atlantic World* (Stanford: Stanford University Press, 2001); Lee Alan Dugatkin, *Mr. Jefferson and the Giant Moose: Natural History in Early America* (Chicago: University of Chicago Press, 2009); Antonello Gerbi, *The Dispute of the New World: The History of a Polemic, 1750–1900,* trans. Jeremy

Moyle (Pittsburgh: University of Pittsburgh Press, 2010 [1955]); Gordon M. Sayre, "Jefferson Takes on Buffon: The Polemic on American Animals in *Notes on the State of Virginia*," *William and Mary Quarterly* 78, no. 1 (January 2021): 79–116.

32. Nicolas Blanquart de Salines, quoted in Buffon, *Natural History, General and Particular,* trans. William Smellie, 3rd ed., vol. 5: 48, 50; Simmons, Sterling, and Watson, "Species and Size-selective Predation by Raccoons (*Procyon lotor*) Preying on Introduced Intertidal Clams," 1061. Buffon's translator William Smellie glossed the pet raccoon of Salines as "he" throughout the text, given the masculine gendering of Buffon's word for raccoon (*un raton*); but Salines revealed at the end of his letter that the raccoon was in fact assigned female (as will become relevant later in the chapter). Although these gendered pronouns are still all-too-human categories, I've amended the pronouns in Smellie's translation. My thanks to Catie Peters and Kenneth Cohen for helping me sort through the gendering in the original account.
33. Edward Long, *The History of Jamaica . . .* (London: T. Lowndes, 1774), vol. 2: 367.
34. Lawson, *A New Voyage to Carolina,* 121; John James Audubon and John Bachman, *The Viviparous Quadrupeds of North America* (New York: V. G. Audubon, 1851), vol. 2: 76. Beginning in 1845, Audubon and Bachman published the first installment of this collaborative multivolume work on American mammals, conveyed (in separate volumes) through lively text and stunning lithographic prints. Bachman authored most of the written text, but the pronoun "we" is used throughout, and a note in the first volume's introduction reads: "For the sake of convenience and uniformity we have written in the plural number, although the facts stated, and the information collected, were obtained at different times by the authors in their individual capacities. . . . Without entering into details of the labours of each in this undertaking, it will be sufficient to add that the history of the habits of our quadrupeds was obtained by both authors, either from personal observation or through the kindness of friends of science, on whose statements full reliance could be placed." See Audubon and Bachman, *The Viviparous Quadrupeds of North America* (New York: J. J. Audubon, 1846), vol. 1: xi. I therefore refer to the authors as Audubon and Bachman throughout this chapter.
35. Audubon and Bachman, *The Viviparous Quadrupeds of North America,* vol. 2: 76; Charles Ball, *Slavery in the United States: A Narrative of the Life and Adventures of Charles Ball . . .* (New York: John S. Taylor, 1837 [1836]), 262. On the evolution of the racial slur, see Justice, "The Racist Coon," in *Raccoon,* 73–109; David R. Roediger, *The Wages of Whiteness: Race and the Making of the American Working Class* (New York: Verso, 2007 [1991]), 98. Gregory Nobles has also shown how Audubon compared enslaved children to raccoons; see Gregory Nobles, *John James Audubon: The Nature of the American Woodsman* (Philadelphia: University of Pennsylvania Press, 2017), 205. On the central role of animals in Ball's narrative, see Thomas G. Andrews, "Beasts of the Southern Wild: Slaveholders, Slaves, and Other Animals in Charles Ball's *Slavery in the United States,*" in *Rendering Nature: Animals, Bodies, Places, Politics,* ed. Marguerite S. Shaffer and Phoebe S. K. Young (Philadelphia: University of Pennsylvania Press, 2015). I am grateful to Tiya Miles for pointing me to the Andrews source.

36. Quoted in Betje Black Klier, *Pavie in the Borderlands: The Journey of Théodore Pavie to Louisiana and Texas, 1829–1830, Including Portions of his* Souvenirs atlantiques (Baton Rouge: Louisiana State University Press, 2000), 217–218; Thomas Bewick, *A General History of Quadrupeds* (Newcastle: Hodgson, Beilby, and Bewick, 1800 [1790]), 4th ed., 279. See also Matthew Mulcahy and Stuart Schwartz, "Nature's Battalions: Insects as Agricultural Pests in the Early Modern Caribbean," *William and Mary Quarterly* 75, no. 3 (July 2018): 433–464.
37. Lawson, *A New Voyage to Carolina*, 121; *The Penny Cyclopaedia of the Society for the Diffusion of Useful Knowledge*, 26 (1843): 57–58; Charles Lyell, *Travels in North America, in the Years 1841–2; with Geological Observations on the United States, Canada, and Nova Scotia* (New York: Wiley and Putnam, 1845), vol. 1: 133.
38. *The Penny Cyclopaedia*, 58. See also Gene Waddell, "Introduction: John Bachman's Works and Life," in *John Bachman: Selected Writings on Science, Race, and Religion*, ed. Gene Waddell (Athens: University of Georgia Press, 2011), 6.
39. Audubon and Bachman, *The Viviparous Quadrupeds of North America*, vol. 2: 78–79.
40. Peter [Pehr] Kalm, *Travels into North America, Containing Its Natural History, and A Circumstantial Account of its Plantations and Agriculture in General*, trans. John Reinhold Forster (London, 1771), vol. 2: 64; De Kay, "Part I. Mammalia," in *Zoology of New-York*, 26; Audubon and Bachman, *The Viviparous Quadrupeds of North America*, vol. 2: 76, 80; Marcy Norton, "The Chicken or the *Iegue:* Human-Animal Relationships and the Columbian Exchange," *American Historical Review* 120, no. 1 (2015): 28–60, esp. 55–56. Although substantial controversy exists over the concept of domestication, many modern-day definitions entail human control of the animal's breeding, a degree of tameness in the animal, and changes in phenotype. However, the term was used more loosely in the eighteenth century. See Harriet Ritvo, "The Domestic Stain, or Maintaining Standards," in "Troubling Species: Care and Belonging in a Relational World," *RCC Perspectives: Transformations in Environment and Society* no. 1 (2017): 19–24. Bachman did have a raccoon in his possession by the time he penned text for *Quadrupeds*, as he wrote in a letter to Audubon's son Victor: "You cannot think how much of my time is taken up in writing about Quadrupeds. . . . We have the Puma—Jaguar—Peccary—American Badger—Grizzly Bear—Black Bear—Fox—Raccoon—Prairie Wolf etc. in a menagerie alive here. I work away at them every two days and find them very serviceable." Letter from John Bachman to Victor Gifford Audubon, January 1846, in *John Bachman: Selected Writings on Science, Race, and Religion*, 345.
41. F. A. Michaux, *Travels to the West of the Alleghany Mountains* . . . (London, 1805), 174; J. D. Schultze, *Bemerkungen über den Waschbären* . . . (Hamburg, 1787), 8; Byrd, *The Dividing Line Histories of William Byrd II of Westover*, 165–166; Sarah Hand Meacham, "Pets, Status, and Slavery in the Late-eighteenth-century Chesapeake," *Journal of Southern History* 77, no. 3 (2011): 521–554. See also *Latrobe's View of America, 1795–1820: Selections from the Watercolors and Sketches*, ed. Edward C. Carter II, John C. Van Horne, and Charles E. Brownell (New Haven: Yale University Press, 1985), 105. I am grateful to Joyce Chaplin for drawing my attention to the Latrobe sketch (figure 7.10).

42. George Shaw, *General Zoology or Systematic Natural History* (London: G. Kearsley, 1800), vol. 1, part 2: 466 ("very cleanly"); Charles Willson Peale, "Lectures, 1800–01," Academy of Natural Sciences, Collection 40, vol. 1; John D. Godman, *American Natural History* (Philadelphia: H. C. Carey and I. Lea, 1826), vol. 1: 164–165.
43. Kathleen M. Brown, *Foul Bodies: Cleanliness in Early America* (New Haven: Yale University Press, 2009), 33–34, 131, and 133.
44. Richard Brookes, *A New and Accurate System of Natural History* . . . (London: J. Newbery, 1763), vol. 1: 237; Edward Turner Bennett, *The Tower Menagerie: Comprising the Natural History of the Animals Contained in that Establishment; with Anecdotes of their Characters and History* (London: Robert Jennings, 1829), 114; Godman, *American Natural History*, vol. 1: 167.
45. Godman, *American Natural History*, vol. 1: 171; Shaw, *General Zoology*, vol. 1, part 2: 465; Goldsmith, *An History of the Earth*, vol. 4: 334–335.
46. Linnaeus, quoted in Blunt, *Linnaeus: The Compleat Naturalist*, 151 ("tobacco pipe"); Linnaeus, "Beskrifning På et Americanskt Diur," trans. Jens Amborg ("coward"); Audubon and Bachman, *The Viviparous Quadrupeds of North America*, vol. 2: 79–80.
47. Godman, *American Natural History*, vol. 1: 169; Audubon and Bachman, *The Viviparous Quadrupeds of North America*, vol. 2: 79 ("mischievous"), 76 ("adroitly").
48. Linnaeus, "Beskrifning På et Americanskt Diur," trans. Jens Amborg.
49. Salines, quoted in Buffon, *Natural History, General and Particular*, trans. William Smellie, 3rd ed., vol. 5: 50; Kalm, *Travels into North America* . . . , trans. John Reinhold Forster (Warrington, U.K.: William Eyres, 1770), vol. 1: 208–209. For the classic treatment of sugar and slavery, see Sidney W. Mintz, *Sweetness and Power: The Place of Sugar in Modern History* (New York: Viking, 1985).
50. Linnaeus, "Beskrifning På et Americanskt Diur," trans. Jens Amborg; Shaw, *General Zoology*, vol. 1, part 2: 466; Salines, quoted in Buffon, *Natural History, General and Particular*, trans. William Smellie, 3rd ed., vol. 5: 50.
51. Charles Hamilton Smith, *Introduction to the Mammalia* (Edinburgh: 1842), 221 ("In captivity"); Bennett, *The Tower Menagerie*, 113–114; Bartram, *Travels*, 341 ("my faithful slave"); William Bartram, "Anecdotes of an American Crow," *Philadelphia Medical and Physical Journal* 1 (1804): 91–95; Ingrid H. Tague, "Companions, Servants, or Slaves?: Considering Animals in Eighteenth-Century Britain," *Studies in Eighteenth-Century Culture* 39 (2010): 111–130. For a fantastic personal and scholarly consideration of crows as historical and contemporary actors—animals that share many qualities, from curiosity to ecological success, with raccoons—see Thom van Dooren, *The Wake of Crows: Living and Dying in Shared Worlds* (New York: Columbia University Press, 2019).
52. *The Penny Cyclopaedia*, vol. 26: 58–59; Godman, *American Natural History*, vol. 1: 169; Salines, quoted in Buffon, *Natural History, General and Particular*, trans. William Smellie, 3rd ed., vol. 5: 49. On anxieties sparked by that ambiguous space between wildness and tameness in nineteenth-century America, see Katherine C. Grier, *Pets in America: A History* (Chapel Hill: University of North Carolina Press, 2006), 183.

53. Salines, quoted in Buffon, *Natural History, General and Particular,* trans. William Smellie, 3rd ed., vol. 5: 52; Immanuel Kant, *The Metaphysics of Morals,* trans. and ed. Mary Gregor (Cambridge: Cambridge University Press, 1996 [1797]), 179. See also Greta LaFleur, *The Natural History of Sexuality in Early America* (Baltimore: Johns Hopkins University Press, 2018), esp. 111–113.
54. Keeping raccoons as quasi-pets and quasi-subjects of experimentation would continue into the twentieth century, as Michael Pettit has shown in his account of the rise and fall of the raccoon as a laboratory animal in behavioral psychology. Many themes from eighteenth-century studies of raccoons persisted in this later debate; as Pettit notes, experimenters believed the "raccoon offered a different model for the animal mind. In terms of mental apparatus, it was the governing role of curiosity that distinguished the raccoon from other mammals." And they viewed the raccoon's curiosity as being "sustained by the species' acute sense of touch, especially in their highly dexterous hands and sensitive nose." See Michael Pettit, "The Problem of Raccoon Intelligence in Behaviourist America," *British Journal for the History of Science* 43, no. 3 (September 2010): 402–403. For a poignant account of animals protesting the cruelties of captivity in another historical context, see Ian J. Miller, *The Nature of the Beasts: Empire and Exhibition at the Tokyo Imperial Zoo* (Berkeley: University of California Press, 2013), esp. 157.
55. On animals and antebellum urbanization, see Catherine McNeur, *Taming Manhattan: Environmental Battles in the Antebellum City* (Cambridge: Harvard University Press, 2014). A number of historians, anthropologists, and ecologists have questioned the term "invasive species" and its colonial history. See Reo and Ogden, "Anishnaabe Aki: An Indigenous Perspective on the Global Threat of Invasive Species," 1443–1452; Stefan Helmreich, "How Scientists Think; About 'Natives,' For Example. A Problem of Taxonomy Among Biologists of Alien Species in Hawaii," *Journal of the Royal Anthropological Institute* 11, no. 1 (March 2005): 107–128. On animal diasporas, see Laura A. Ogden, "The Beaver Diaspora: A Thought Experiment," *Environmental Humanities* 10, no. 1 (2018): 63–85. On declension narratives and environmental history, see William Cronon, "A Place for Stories: Nature, History, and Narrative," *Journal of American History* 78, no. 4 (March 1992): 1347–1376.
56. Suzanne E. MacDonald and Sarah Ritvo, "Comparative Cognition Outside the Laboratory," *Comparative Cognition & Behavior Reviews* 11 (2016): 49–62 ("selecting," 55); MacDonald, quoted in Liz Langley, "Raccoons Pass Famous Intelligence Test—By Upending It," *National Geographic* (October 21, 2017), https://news.nationalgeographic.com/2017/10/animals-intelligence-raccoons-birds-aesops/ (accessed April 15, 2019); Suzanne MacDonald, "Urban Raccoons in the Greater Toronto Area: Are We Building a Smarter Raccoon?" Reinventing the City: A Workshop on Habitecture for Wildlife, University of Toronto and York University, www.youtube.com/watch?v=FLwHKirq-SY (accessed September 22, 2022).
57. Buffon, *Natural History, General and Particular,* trans. William Smellie, 2nd ed., vol. 2: 364 ("were endowed"), vol. 4: 73 ("talents").
58. I part ways with scholars who do not view animals as capable of change over time

in the historical sense. Anthropologists David Premack and Ann James Premack, for instance, have argued that "animals have not undergone significant change while remaining biologically stable," even over the course of millions of years—though recent findings in animal behavior suggest otherwise. See David Premack and Ann James Premack, "Why Animals Have Neither Culture Nor History," in *Companion Encyclopedia of Anthropology: Humanity, Culture and Social Life,* ed. Tim Ingold (London: Routledge, 1994), 350.

59. Buffon, *Natural History, General and Particular,* trans. William Smellie, 2nd ed., vol. 4: 73.

CHAPTER 8. TOUCHING THE PAST

1. On the Anthropocene and its critics, see Dipesh Chakrabarty, "The Climate of History: Four Theses," *Critical Inquiry* 35, no. 2 (Winter 2009): 197–222; Heather Davis and Zoe Todd, "On the Importance of a Date, or Decolonizing the Anthropocene," *ACME: An International Journal for Critical Geographies* 16, no. 4 (2017): 761–780; Donna Haraway, *Staying with the Trouble: Making Kin in the Chthulucene* (Durham, N.C.: Duke University Press, 2016); Timothy James LeCain, "Against the Anthropocene: A Neo-Materialist Perspective," *International Journal for History, Culture, and Modernity* 3, no. 1 (2015): 1–28; J. R. McNeill, "Introductory Remarks: The Anthropocene and the Eighteenth Century," *Eighteenth-Century Studies* 49, no. 2 (2016): 117–128.
2. The official population listed by the IUCN is currently 192 "mature" individuals, last evaluated in 2016. See A. D. Cuarón, P. C. de Grammont, and K. McFadden, *Procyon pygmaeus. The IUCN Red List of Threatened Species* (2016): e.T18267A45201913, https://dx.doi.org/10.2305/IUCN.UK.2016-1.RLTS.T18267A45201913.en (accessed October 5, 2022). For the SEMARNAT designation, see Alejandro Flores-Manzanero et al., "Conservation Genetics of Two Critically Endangered Island Dwarf Carnivores," *Conservation Genetics* 23 (2022): 35–49. On the contested Guadeloupe raccoon—not considered to be a separate species from the common raccoon, at least not by current biologists—which has become another Caribbean island mascot, see Simon Adler, "Stranger in Paradise," *Radiolab,* podcast audio, January 27, 2017, www.wnycstudios.org/podcasts/radiolab/articles/stanger-paradise (accessed October 5, 2022).
3. For an ethnography of Cozumel's cruise ship industry, see Christine Preble, "Imperial Consumption: Cruise Ship Tourism and Cozumel, Mexico" (PhD dissertation, University at Albany, State University of New York, 2014).
4. Melissa Groo, "How to Photograph Wildlife Ethically," *National Geographic* (July 31, 2019), www.nationalgeographic.com/animals/2019/07/ethical-wildlife-photography/ (accessed October 6, 2022).
5. See Cronon, "The Trouble with Wilderness"; "From Feed the Birds to Do Not Feed the Animals," https://animalfeeding.org/ (accessed October 6, 2022).
6. Katherine W. McFadden et al., "Feeding Habits of Endangered Pygmy Raccoons (*Procyon pygmaeus*) Based on Stable Isotope and Fecal Analysis," *Journal of Mammalogy* 87, no. 3 (2006): 501–509.

7. Alfredo D. Cuarón et al., "The Status of Dwarf Carnivores on Cozumel Island, Mexico," *Biodiversity and Conservation* 13 (2004): 321 ("remarkably reduced").
8. McFadden et al., "Feeding Habits of Endangered Pygmy Raccoons"; Flores-Manzanero et al., "Conservation Genetics of Two Critically Endangered Island Dwarf Carnivores," esp. 37 and 45 ("Therefore").
9. Lulu Miller, "Animal Planet," *Orion* (Spring 2020), https://orionmagazine.org/article/animal-planet/ (accessed October 7, 2022).
10. Katherine W. McFadden, "Vulnerable Island Carnivores: The Endangered Endemic Dwarf Procyonids from Cozumel Island," *Biodiversity and Conservation* 19, no. 2 (February 2009): 491–502; Flores-Manzanero et al., "Conservation Genetics of Two Critically Endangered Island Dwarf Carnivores."
11. Vivien Louppe et al., "New Insights on the Geographical Origins of the Caribbean Raccoons," *Journal of Zoological Systematics and Evolutionary Research* 58, no. 4 (November 2020): 1303–1322; Flores-Manzanero et al., "Conservation Genetics of Two Critically Endangered Island Dwarf Carnivores."

EPILOGUE

1. Kent E. Carpenter and Rodolfo B. Reyes, "*Somniosus microcephalus* (Bloch & Schneider, 1801)," FishBase, ed. R. Froese and D. Pauly, 2022, www.fishbase.de/summary/138 (accessed September 25, 2022).
2. Julius Nielsen et al., "Eye Lens Radiocarbon Reveals Centuries of Longevity in the Greenland Shark (*Somniosus microcephalus*)," *Science* 353, no. 6300 (August 2016): 702–704; M. E. Blochii and Jo. Gottlob Schneider, *Systema Ichthyologiae* . . . (Berlin: 1801), 135. After Bloch died in 1799, Schneider published the *Systema*. On the lost specimen: Edda Aßel, email conversation with the author, April 4, 2022. For the last known "sighting" of Bloch's specimen, see J. Müller and J. Henle, *Systematische Beschreibung der Plagiostomen* (Berlin: 1841), 94.
3. E. Brendan Roark et al., "Extreme Longevity in Proteinaceous Deep-sea Corals," *Proceedings of the National Academy of Sciences* 106, no. 13 (March 2009): 5204–5208.
4. John Burroughs, "Real and Sham Natural History," *Atlantic Monthly* 91 (March 1903): 305. For the definitive account of the controversy, see Ralph H. Lutts, *The Nature Fakers: Wildlife, Science, and Sentiment* (Charlottesville: University Press of Virginia, 2001 [1990]).
5. Richard O. Prum, quoted in Ferris Jabr, "How Beauty Is Making Scientists Rethink Evolution," *New York Times Magazine,* January 9, 2019, www.nytimes.com/2019/01/09/magazine/beauty-evolution-animal.html (accessed October 7, 2022); Richard O. Prum, *The Evolution of Beauty: How Darwin's Forgotten Theory of Mate Choice Shapes the Animal World* (New York: Doubleday, 2017), 7.
6. On the gut-brain connection in recent microbiome research, see Cassandra Willyard, "How Gut Microbes Could Drive Brain Disorders," *Nature* News Feature, February 3, 2021, www.nature.com/articles/d41586-021-00260-3#correction-0 (accessed August 24, 2022).
7. Linné, *A General System of Nature,* vol. 1: 9 (quotations). For the original Latin

description of *Homo sapiens* in the tenth edition of the *Systema Naturae,* see Linnaeus, *Systema naturae per regna tria naturae,* 10th ed., vol. 1: 20–22. There is some disagreement as to whether Linnaeus's description of the European variety as "*Regitur* ritibus" (21) should be translated as "governed by laws" or "governed by rites." The 1802 translation I cite uses "laws." On the clear hierarchy implied despite Linnaeus's murky intent with that wording, see James Q. Whitman, "The World Historical Significance of European Legal History: An Interim Report," in *The Oxford Handbook of European Legal History,* ed. Heikki Pihlajamäki, Markus D. Dubber, and Mark Godfrey (Oxford: Oxford University Press, 2018), 8. For one summary of the anthropological debate between Marshall Sahlins and Gananath Obeyesekere regarding Cook's deification (or lack thereof), see K. R. Howe, "Review Article: The Making of Cook's Death," *Journal of Pacific History* 31, no. 1 (June 1996): 108–118.

INDEX